tredition

tredition was established in 2006 by Sandra Latusseck and Soenke Schulz. Based in Hamburg, Germany, tredition offers publishing solutions to authors and publishing houses, combined with worldwide distribution of printed and digital book content. tredition is uniquely positioned to enable authors and publishing houses to create books on their own terms and without conventional manufacturing risks.

For more information please visit: www.tredition.com

TREDITION CLASSICS

This book is part of the TREDITION CLASSICS series. The creators of this series are united by passion for literature and driven by the intention of making all public domain books available in printed format again - worldwide. Most TREDITION CLASSICS titles have been out of print and off the bookstore shelves for decades. At tredition we believe that a great book never goes out of style and that its value is eternal. Several mostly non-profit literature projects provide content to tredition. To support their good work, tredition donates a portion of the proceeds from each sold copy. As a reader of a TREDITION CLASSICS book, you support our mission to save many of the amazing works of world literature from oblivion. See all available books at www.tredition.com.

 Project Gutenberg

The content for this book has been graciously provided by Project Gutenberg. Project Gutenberg is a non-profit organization founded by Michael Hart in 1971 at the University of Illinois. The mission of Project Gutenberg is simple: To encourage the creation and distribution of eBooks. Project Gutenberg is the first and largest collection of public domain eBooks.

Birds of Guernsey (1879) And the Neighbouring Islands: Alderney, Sark, Jethou, Herm, Being a Small Contribution to the Ornitholony of the Channel Islands

Cecil Smith

Imprint

This book is part of TREDITION CLASSICS

Author: Cecil Smith
Cover design: Buchgut, Berlin – Germany

Publisher: tredition GmbH, Hamburg - Germany
ISBN: 978-3-8424-7590-8

www.tredition.com
www.tredition.de

Copyright:
The content of this book is sourced from the public domain.

The intention of the TREDITION CLASSICS series is to make world literature in the public domain available in printed format. Literary enthusiasts and organizations, such as Project Gutenberg, worldwide have scanned and digitally edited the original texts. tredition has subsequently formatted and redesigned the content into a modern reading layout. Therefore, we cannot guarantee the exact reproduction of the original format of a particular historic edition. Please also note that no modifications have been made to the spelling, therefore it may differ from the orthography used today.

CONTENTS.

PREFACE.
BIRDS OF GUERNSEY.
ENDNOTES.

PREFACE.

Though perhaps not possessing the interest to the ornithologist which Lundy Island (the only breeding-place of the Gannet in the South-West of England) or the Scilly Islands possess, or being able to produce the long list of birds which the indefatigable Mr. Gdetke has been able to do for his little island, Heligoland, the avifauna of Guernsey and the neighbouring islands is by no means devoid of interest; and as little has hitherto been published about the Birds of Guernsey and the neighbouring islands, except in a few occasional papers published by Miss C.B. Carey, Mr. Harvie Browne, myself, and a few others, in the pages of the 'Zoologist,' I make no excuse for publishing this list of the birds, which, as an occasional visitor to the Channel Islands for now some thirty years, have in some way been brought to my notice as occurring in these Islands either as residents, migrants, or occasional visitants.

Channel Island specimens of several of the rarer birds mentioned, as well as of the commoner ones, are in my own collection; and others I have seen either in the flesh or only recently skinned in the bird-stuffers' shops. For a few, of course, I have been obliged to rely on the evidence of others; some of these may appear, perhaps, rather questionable,—as, for instance, the Osprey,—but I have always given what evidence I have been able to collect in each case; and where evidence of the occurrence was altogether wanting, I have thought it better to omit all mention of the bird, though its occasional occurrence may seem possible.

I have confined myself in this list to the Birds of Guernsey and the neighbouring islands—Sark, Alderney, Jethou and Herm; in fact to the islands included in the Bailiwick of Guernsey. I have done this as I have had no opportunity of personally studying the birds of Jersey, only having been in that island once some years ago, and then only for a short time, and not because I think a notice of the birds of Jersey would have been devoid of interest, though whether it would have added many to my list maybe doubtful. Professor Ansted's list, included in his large and very interesting work on the

Channel Islands, is hitherto the only attempt at a regular list of the Birds of the Channel Islands; but as he, though great as a geologist, is no ornithologist, he was obliged to rely in a great measure on information received from others, and this apparently was not always very reliable, and he does not appear to have taken much trouble to sift the evidence given to him. Professor Ansted himself states that his list is necessarily imperfect, as he received little or no information from some of the Islands; in fact, Guernsey and Sark appear to be the only two from which much information had been received. This is to be regretted, as it has made the notice of the distribution of the various birds through the Islands, which he has denoted by the letters *a, e, i, o, u* [1] appended to the name of each bird, necessarily faulty. The ornithological notes, however, supplied by Mr. Gallienne are of considerable interest, and are generally pretty reliable. It is rather remarkable, however, that Professor Ansted has not always paid attention to these notes in marking the distribution of the birds through the various Islands.

No doubt many of the birds included in Professor Ansted's list were included merely on the authority of specimens in the museum of the Mechanics' Institute, which at one time was a pretty good one; and had sufficient care been taken to label the various specimens correctly as to place and date, especially distinguishing local specimens from foreign ones, of which there were a good many, would have been a very interesting and useful local museum; as it is, the interest of this museum is considerably deteriorated. Some of the birds in the museum are confessedly foreign, having been brought from various parts of the world by Guernsey men, who when abroad remembered the museum in their own Island, and brought home specimens for it. Others, as Mr. Gallienne, who during his life took much interest in the museum, himself told me had been purchased from various bird-stuffers, especially from one in Jersey; and no questions were asked as to whether the specimens bought were local or set-up from skins obtained from the Continent or England. Amongst those so obtained may probably be classed the Blue-throated Warblers, included in Professor Ansted's list and marked as Jersey (these Mr. Gallienne himself told me he believed to be Continental and not genuine Channel Island specimens), the

Great Sedge Warbler, the Meadow Bunting, the Green Woodpecker, and perhaps a few others.

This museum, partly from want of interest being taken in it and partly from want of money, has never had a very good room, and has been shuffled and moved about from one place to another, and consequently several birds really valuable, as they could be proved to be genuine Channel Island specimens, have been lost and destroyed; in fact, had it not been for the care and energy of Miss C.B. Carey, who took great pains to preserve what she found remaining of the collection, and place it in some sort of order, distinguishing by a different coloured label those specimens which could be proved to be Channel Island (in doing this she worked very hard, and received very little thanks or encouragement, but on the contrary met with a considerable amount of genuine obstructiveness), the whole of the specimens in the museum would undoubtedly have been lost; as it is, a good many valuable local specimens — valuable as being still capable of being proved to be genuine Channel Island specimens — have been preserved, and a good nucleus kept for the foundation of a new museum, should interest in the subject revive and the local authorities be disposed to assist in its formation. In my notices of each bird I have mentioned whether there is a specimen in the museum, and also whether it is included in Professor Ansted's list, and if so in which of the Islands he has marked it as occurring.

No doubt the Ornithology of the Channel Islands, as is the case in many counties of England, has been considerably changed by drainage works, improved cultivation, and road-making; much alteration of this sort I can see has taken place during the thirty years which I have known the Islands as an occasional visitor. But Mr. MacCulloch, who has been resident in the Islands for a much longer period — in fact, he has told me nearly double — has very kindly supplied me with the following very interesting note on the various changes which have taken place in Guernsey during the long period he has lived in that island; he says, "I can well recollect the cutting of most of the main roads, and the improvement, still going on, of the smaller ones. It was about the beginning of this century that the works for reclaiming the Braye du Valle were undertaken; before that time the Clos du Valle [2] was separated from the mainland by an arm of the sea, left dry at low water, extending from St. Samson's to the Vale Church. This was bordered by salt marshes only, covered occasionally at spring tides by the sea, some of which extended pretty far inland. The meadows adjoining were very imperfectly drained, as indeed some still are, and covered with reeds and rushes, forming excellent shelter for many species of aquatic birds. Now, as you know, by far the greater part of the land is well cultivated and thickly covered with habitations. The old roads were everywhere enclosed between high hedges, on which were planted rows of elms; and the same kind of hedge divided the fields and tenements. Every house, too, in those days had its orchard, cider being then universally drunk; and the hill-sides and cliffs were covered with furze brakes, as in all country houses they baked their own bread and required the furze for fuel. Now all that is changed. The meadows are drained and planted with brocoli for the early London market, to be replaced by a crop of potatoes at the end of the summer. The trees are cut down to let in the sun. Since the people have taken to gin-drinking, cider is out of favour and the orchards destroyed. The hedges are levelled to gain a few perches of ground, and replaced in many places by stone walls; the furze brakes rooted up, and the whole aspect and nature of the country changed. Is it to be wondered at that those kinds of birds that love shelter and quiet have deserted us? You know, too, how every bird—from the Wren to the Eagle—is popped at as soon as it shows itself, in places where there are no game laws and every man allowed to carry a gun."

This interesting description of the changes—agricultural and otherwise—which have taken place in the Islands, especially Guernsey, during the last fifty or sixty years (for which I have to offer Mr. MacCulloch my best thanks), gives a very good general idea of many of the alterations that have taken place in the face of the country during the period above mentioned; but does not by any means exhaust them, as no mention is made of the immense increase of orchard-houses in all parts of Guernsey, which has been so great that I may fairly say that within the last few years miles of glass-houses have been built in Guernsey alone: these have been built mostly for the purpose of growing grapes for the London market. These orchard-houses have, to a certain extent, taken the place of ordinary orchards and gardens, which have been rooted up and destroyed to make place for this enormous extent of glass. But what appeared to me to have made the greatest change, and has probably had more effect on the Ornithology of the Island, especially of that part known as the Vale, is the enormous number of granite quarries which are being worked there (luckily the beautiful cliffs have hitherto escaped the granite in those parts, probably not being so good); but in the Vale from St. Samson's to Fort Doyle, and from there to the Vale Church, with the exception of L'Ancresse Common itself, which has hitherto escaped, the whole face of the country is changed by quarry works and covered with small windmills used for pumping the water from the quarries. These quarry works and the extra population brought by them into the Island, all of whom carry guns and shoot everything that is fit to eat or is likely to fetch a few "doubles" in the market, have done a good deal to thin the birds in that part of the Islands, especially such as are in any way fit for sale or food, and probably have done more to make a change in the Ornithology of that part of the Island than all the agricultural changes mentioned by Mr. MacCulloch. Indeed, I am rather sceptical as to the agricultural changes above described having produced so much change in the avifauna of the Islands during the last fifty years as Mr. MacCulloch appears to think; there is still a great deal of undrained or badly drained land in the Island—especially about the Vale, the Grand Mare and L'Eree—which might still afford a home for Moorhens, Water Rails, and even Bitterns, and all that class of wading birds which delight in swampy land and reed beds. Though no doubt, as Mr. MacCulloch said, many orchards have been destroyed to make room for more profitable crops or for orchard-houses, still there are many orchards left in the Island. I think, however, many, if not all the cherry orchards (amongst which the Golden Orioles apparently at one time luxuriated) are gone. There is also still a great deal of hedgerow timber, none of it indeed very large, but in places very thick; in fact, I could point out miles of hedges in Guernsey where the trees, mostly elm, grow so thick together that it would be nearly impossible to pick out a place where one could squeeze one's horse between the trees without rubbing one's knees on one side or the other, probably on both, against them, if one found it necessary to ride across the country. True, on a great extent of the higher part of the Island, all along on both sides of what is known as the Forest Road, there is little or no hedgerow timber, the fields here being divided by low banks with furze growing on the top of them. Furze brakes also are still numerous, the whole of the flat land on the top of the cliffs and the steep valleys and slopes down to the sea on the south and east side of the Island, from Fermain Bay to Pleimont, being almost uninterrupted wild land covered with heather, furze, and bracken; besides this wild furze land, there are several thick furze brakes inland in different parts of the Island. All these places seem to me to have remained almost without change for years. The furze, however, never grows very high, as it is cut every few years for fuel; in consequence of this, however, it is more beautiful in blooming in the spring than if it had been allowed several years' growth, covering the whole face of the ground above the cliffs like a brilliant yellow carpet; but being kept so short, it is not perhaps so convenient for nesting purposes as if it was allowed a longer growth.

The Guernsey Bird Act, which applies to all the Islands in the Bailiwick, and has been in force for some few years, seems to me to have had little effect on the numbers of the sea-birds of the district, though it includes the eggs as well as the birds, except perhaps to increase the number of Herring Gulls and Shags (which were always sufficiently numerous) in their old breeding-stations, and perhaps to have added a few new breeding-stations. These two birds scarcely needed the protection afforded by the Act, as their nests are placed amongst very inaccessible rocks where very few nests can be reached without the aid of a rope, and consequently but little damage was done beyond a few young birds being shot soon after they had left the nest while they were flappers, and the numbers were fully kept up; other birds, however, included in the Act, and not breeding in quite such inaccessible places, seem to gain but little advantage from it, as nests of the Lesser Black-backed Gulls, Terns, Oystercatchers and Puffins are ruthlessly robbed in a way that bids fair before long to exterminate all four species as breeding birds; perhaps, also, the increase in the number of Herring Gulls does something to diminish the numbers of other breeding species, especially the Lesser Black-backs, as Herring Gulls are great robbers both of eggs and young birds. The Act itself, after reciting that "le nombre des oiseaux de mer sur les cttes des Isles de cet

Bailliage a considerablement diminui depuis plusieurs annies; que les dits oiseaux sont utiles aux pjcheurs, en ce qu'ils indiquent les parages ou les poissons se trouvent; que les dits oiseaux sont utiles aux marins en ce qu'ils annoncent pendant la durie des brouillards la proximite des rochers," goes on to enact as follows: — "Il est difendu de prendre, enlever ou ditruire les ceufs des oiseaux de mer dans toute l'entendue de la jurisdiction de cette isle, sur la peine d'une amende qui ne sera pas moindre de sept livres tournois et n'excidera pas trente livres tournois." [3] Sec. 2 enacts, "Depuis ce jour [4] au 15 Octobre prochain, il est difendu de tuer, blesser, prendre ou chasser les oiseaux de mer dans toute l'entendue de la jurisdiction de cette isle." Sec. 3, "Ceux qui depuis ce jour au 15 Octobre prochain auront iti trouvis en possession d'un oiseau de mer ricemment tui, blessi ou pris, ou qui auront iti trouvis en possession de plumage frais appartenant d'un oiseau de mer seront censis avoir tui, blessi ou pris tel oiseau de mer sauf h eux de prouver le contraire. Pareillement ceux qui depuis ce jour au 15 Octobre prochain auront iti trouvis en possession d'un oeuf de l'annee d'un oiseau de mer seront censis avoir pris et enleve le dit oeuf sauf ` eux de prouver le contraire." The penalty in each case is the same as in Section 1. Section 4 contains the list of the oiseaux de mer which come under the protection of the Act, which is as follows: — Les Mauves Mouettes, Pingouins, Guillemots, Cormorans, Barbelotes, Hirondelles de mer, Pies-marants, Petrel, Plongeons, Grebes, Puffins, Dotterells, Alouettes de mer, Toumpierres, Gannets, Courlis et Martin pjcheur.

As far as the eggs of many of the species actually breeding in the Islands are concerned, this Act seems to be a dead letter: the only birds of any size whose eggs are not regularly robbed are the Herring Gulls and Shags, and they take sufficient care of themselves; were the Act strictly enforced it would probably be found that there would be — as would be the case in England — a good deal of opposition to this part of it, which would greatly interfere with what appears to be a considerable article of food with many of the population. Probably the only compromise which would work, and could be rigidly enforced, would be to fix a later date for the protection of the eggs — say as late as the 15th June; this would allow those who wanted to rob the eggs for food to take the earlier layings, and the birds would be able to bring up their second or third broods in peace; and probably the fishermen and others, who use the eggs as an article of consumption, would be glad to assist in carrying out such an Act as this, as they would soon find the birds increase so much that they would be able to take as many eggs by the middle of June as they do now in the whole year, especially the Black-back Gulls and the Puffins, which are the birds mostly robbed, — the latter of which are certainly decreasing considerably in numbers in consequence.

This plan is successfully carried out by many private owners of the large breeding-stations of the Gannets, Eider Duck, and other sea-birds in the north of England and Scotland. Of course, it must not be supposed that all the birds mentioned in the Act whose eggs are protected breed in the Islands, or anywhere within ten or fifteen degrees of latitude of the Islands; in fact, a great many of them are not there at all during the breeding-season, except perhaps an occasional wounded bird which has been unable to join its companions on their migratory journey, or a few non-breeding stragglers.

It has often struck me that a small but rigidly collected and enforced gun-tax would be a more efficacious protection — not only to the oiseaux de mer, but also to the inland birds, many of which are quite as much in want of protection though not included in the Act — than the Sea-bird Protection Act is. I am glad to see that there is some chance of this being carried out, for, while this work was going through the press, I see by the newspaper ('Gazette Officielle de Guernsey' for the 26th March, 1879) that the Bailiff had then just issued a *Billet d'Etat* which contained a "Projet de loi" on the subject, to be submitted to the States at their next meeting; and in concluding its comments on this *Projet de loi* the Gazette says, "Il n'est que juste en fait que ceux qui veulent se lier au plaisir de la chasse paient pour cette fantaisie et que par ce moyen le trop grand nombre de nos chasseurs maladroits et inexpirimentes se voit riduit au grand avantage de nos fermiers et de nos promeneurs;" and probably also to the advantage of the chasseurs themselves.

In regard to the nomenclature, I have done the best I can to follow the rule laid down by the British Association; but not living in London, and consequently not having access to a sufficiently large ornithological library to enable me to search out the various synonyms for myself and ascertain the exact dates, I have therefore been obliged to rely on the best authorities whose works I possess, and accept the name given by them. In doing this, I have no doubt I have been quite as correct as I should have been had I waded through the various authors who have written on the subject, as I have invariably accepted the name adopted by Professor Newton in his edition of Yarrell, and by Mr. Dresser in his 'Birds of Europe', as far as these works are yet complete: for the birds not yet included in either I have for the most part taken the scientific names from Mr. Howard Saunders's 'Catalogue des oiseaux du midi de L'Espagne,' published in the 'Proceedings' of the Sociiti Zoologique de France; and for the names of the Gulls and Terns I have entirely followed Mr. Howard Saunders's papers on those birds published in the 'Proceedings' of our own Zoological Society, for permission to use which, and for other assistance, — especially in egg-hunting, — I have to give him my best thanks.

As French is so much spoken in Guernsey and the other Islands included in my district, I have (wherever I have been able to ascertain it) given the French name of each bird, as it may be better known to my Guernsey readers than either the English or the scientific name. I have also, where there is one and I have been able to ascertain it, mentioned the local name in the course of my notes on each bird.

It now only remains to give my best thanks to the various friends who have assisted me, especially to Mr. MacCulloch, who, though he says he is no naturalist, has supplied me with various very interesting notes, which he has taken from time to time of ornithological events which have occurred in Guernsey, and from which I have drawn rather largely; and I have, also, again to thank him for the interesting accounts he has given me of the various changes — agricultural and otherwise — which have taken place during his memory, and which may have had some effect on the ornithology of the Islands, especially of Guernsey.

My thanks are also due to Col. L'Estrange for the assistance he has given me in egg-hunting, and also to Captain Hubback for his notes from Alderney during the times he was quartered there.

BIRDS OF GUERNSEY.

1. WHITE-TAILED EAGLE. *Haliaeetus albicilla*, Linnsaeus. French, "Aigle pygarque," "Pygarque ordinaire."—The White-tailed Eagle is an occasional but by no means uncommon visitant to all the Islands. I have seen specimens from Alderney, Guernsey, and Herm, and have heard of its having been killed in Sark more than once. It usually occurs in the autumn, and, as a rule, has a very short lease of life after its arrival in the Islands, which is not to be wondered at, as it is considered, and no doubt is, mischievous both to sheep and poultry; and in so thickly populated a country, where every one carries a gun, a large bird like the White-tailed Eagle can hardly escape notice and consequent destruction for any length of time. It might, however, if unmolested, occasionally remain throughout the winter, and probably sometimes wanders to the Islands at that time, as Mr. Harvie Brown records ('Zoologist' for 1869, p. 1591) one as having been killed, poisoned by strychnine, in Herm in the month of January. This was, no doubt, a late winter visitant, as it is hardly possible that the bird can have escaped for so long a time, as it would have done had it visited the Islands at its usual time, October or November. All the Channel Island specimens of the White-tailed Eagle which I have seen have been young birds of the first or second year, in the immature plumage in which the bird is known as the Sea Eagle of Bewick, and in which it is occasionally mistaken for the Golden Eagle, which bird has never, I believe, occurred in the Islands. Of course in the adult plumage, when this bird has its white tail and head, no such mistake could occur, but in the immature plumage in which the bird usually makes its appearance such a mistake does occasionally happen, and afterwards it becomes difficult to convince the owner that he has not a Golden Eagle; in fact he usually feels rather insulted when told of his mistake, and ignores all suggestions of anything like an infallible test, so it may be as well to mention that the birds may be distinguished in any state of plumage and at any age by the tarsus, which in the White-tailed Eagle is bare of feathers and in the Golden Eagle is feathered to the junction of the toes. I have one in my possession shot at Bordeaux

harbour on the 14th of November, 1871, and I saw one in the flesh at Mr. Couch's, the bird-stuffer, which had been shot at Alderney on the 2nd of November in the same year; and Mr. MacCulloch writes to me that one was wounded and taken alive in the parish of the Forest in Guernsey in 1845. It was said to be one of a pair, and he adds—"I have known several instances of its appearance since both here (Guernsey) and in Herm," but unluckily he gives no dates and could not remember at what time of year any of the occurrences he had noted had taken place. This is to be regretted, as although the bird occurs almost every autumn—indeed, so frequently as to render mention of further instances of its occurrence at that time of year unnecessary—its occurrence in the spring is rare, and some of those noted by Mr. MacCulloch might have been at that time of year. As it is, I only know of one spring occurrence, and that was reported to me by Mr. Couch as having taken place at Herm on the 23rd of March, 1877.

The White-tailed Eagle is included in Professor Ansted's list, but its range in the Islands is restricted to Guernsey. There is one in the museum, probably killed in Guernsey, in the plumage in which the Channel Island specimens usually occur, but no note is given as to locality or date.

2. OSPREY. *Pandion halioeetus*, Linnaeus. French, "Balbusard."—I have never met with the Osprey myself in the Channel Islands, nor have I, as far as I remember, seen a Channel Island specimen. I include it, however, on the authority of a note kindly sent to me by Mr. MacCulloch, who says:—"An Osprey was shot at St. Samsons, in Guernsey, on the 29th of October, 1868. I cannot, however, say whether at the time it was examined by a competent naturalist, and as both the Osprey and the White-tailed Eagle are fishers, a mistake may have been made in naming it." Of course such a mistake as suggested is possible, but as the Guernsey fishermen and gunners, especially the St. Samsons men, are well acquainted with the White-tailed Eagle, I should not think it probable that the mistake had been made. The bird, however, cannot be considered at all common in the Islands; there is no specimen in the Guernsey Museum, and

Mr. Couch has never mentioned to me having had one through his hands, or recorded it in the 'Zoologist,' as he would have done had he had one; neither does Mrs. Jago (late Miss Cumber), who used to do a good deal of stuffing in Guernsey about thirty years ago, remember having had one through her hands. There can be no reason, however, why it should not occasionally occur in the Islands, as it does so both on the French and English side of the Channel. The wonder rather is that it is so rare as it appears to be.

The Osprey, however, is mentioned in Professor Ansted's list, and only marked as occurring in Guernsey.

3. GREENLAND FALCON. *Falco candicans*, Gmelin. — I was much surprised on my last visit to Alderney, on the 27th of June, 1878, on going into a small carpenter's shop in the town, whose owner, besides being a carpenter, is also an amateur bird-stuffer, though of the roughest description, to find, amongst the dust of his shop, not only the Purple Heron, which I went especially to see, and which is mentioned afterwards, but a young Greenland Falcon which he informed me had been shot in that island about eighteen months ago. This statement was afterwards confirmed by the person who shot the bird, who was sent for and came in whilst I was still in the shop. Unfortunately, neither the carpenter nor his friend who shot the bird had made any note of the date, and could only remember that the one had shot the bird in that Island about eighteen months ago and the other had stuffed it immediately after. This would bring it to the winter of 1876-77, or, more probably, the late autumn of 1876. In the course of conversation it appeared to me that the Snow Falcon — as they called this bird — was not entirely unknown to the carpenter or his friend, though neither could remember at the time another instance of one having been killed in that Island. It is, however, by no means improbable that either this species or the next mentioned, or both, may have occurred in the Islands before, as Professor Ansted, though he gives no date or locality, includes the Gyr Falcon in his list of Channel Island birds. As all three of the large northern white Falcons were at one time included under the name of Gyr Falcons, and, as Professor Ansted gives no description

of the bird mentioned by him, it is impossible to say to which species he alluded. We may fairly conclude, however, that it was either the present species or the Iceland Falcon, as it could hardly have been the darker and less wandering species, the Norway Falcon, the true Gyr Falcon of falconers, *Falco gyrfalco* of Linnaeus, which does not wander so far from its native home, and has never yet, as far as is at present known, occurred in any part of the British Islands, and certainly not so far south as the Channel Islands. This latter, indeed, is an extremely southern latitude for either the Greenland or Iceland Falcon, the next being in Cornwall, from which county both species have been recorded by Mr. Rodd. Neither species, however, is recorded as having occurred in any of the neighbouring parts of France.

4. ICELAND FALCON. *Falco islandus*, Gmelin.—An Iceland Falcon was killed on the little Island of Herm on the 11th of April, 1876, where it had been seen about for some time, by the gamekeeper. It had another similar bird in company with it, and probably the pair were living very well upon the game-birds which had been imported and preserved in that island, as the keeper saw them kill more than one Pheasant before he shot this bird. The other fortunately escaped. The bird which was killed is now in my possession, and is a fully adult Iceland Falcon, and Mr. Couch, the bird-stuffer who skinned it, informed me a male by dissection. Though to a certain extent I have profited by it, so far as to have the only Channel Island example of the Iceland Falcon in my possession, I cannot help regretting that this bird was killed by the keeper, as it seems to me not impossible that the two birds being together in the island so late as the 11th of April, and certainly one, probably both, being adult, and there being plenty of food for them, might, if unmolested, have bred in the island. Perhaps, however, this is too much to have expected so far from their proper home. It would, however, have been interesting to know how late the birds would have remained before returning to their northern home; but the breeding-season for the Pheasants was beginning, and this was enough for the keeper, as he had actually seen two or three Pheasants—some hens—killed before he shot the Falcon. As these Falcons can only be considered very

rare accidental visitors to the Islands, it may be interesting to some of my readers to mention that they may distinguish them easily by colour, the Greenland, *Falco candicans*, being always the most white, and the Norway bird — the Gyr Falcon of falconers — being the darkest, the Iceland Falcon (the present species) being intermediate. This is generally a good guide at all ages, but occasionally there may be some difficulty in distinguishing young birds, especially as between the Iceland and the Norway Falcon. In a doubtful case in the Channel Islands, however, it would always be safer to consider the bird an Iceland rather than a Norway Falcon.

5. PEREGRINE FALCON. *Falco peregrinus*, Tunstall. French, "Faucon phlerin." — The Peregrine can now, I think, only be considered an autumnal visitant to the Islands, though, if not shot or otherwise destroyed, it would, no doubt, remain throughout the winter, and might perhaps have been resident, as Mr. MacCulloch sends me a note of one killed in Herm in December. All the Channel Island specimens I have seen have been young birds of the year, and generally killed in October or November. Adult birds, no doubt, occasionally occur, but they are comparatively rare, and it certainly does not breed anywhere in the Islands at present, though I see no reason why it should not have done so in former times, as there are many places well suited to it, and a constant supply of sea-birds for food. Mr. MacCulloch also seems to be of opinion that the Peregrine formerly bred in the Islands, as he says, speaking, however, of the *Falconidae* generally, "There must have been a time when some of the species were permanent residents, for the high pyramidal rock south of the little Island of Jethou bears the name of 'La Fauconnihre,' evidently denoting that it must have been a favourite resort of these birds, and there are other rocks with the same name." Certainly the rock here mentioned looks much like a place that would be selected by the Peregrine for breeding purposes, but that must have been before the days of excursion steamers once or twice a week to Jethou and Herm. Occasionally a young Peregrine is made to do duty as a Lanner, and is recorded in the local papers accordingly (see 'Star' for November 11th, 1876, copying, however, a Jersey paper), but in spite of these occasional notes there is no satisfactory

reason for supposing that the true Lanner has ever occurred in either of the Islands. The birds, however, certainly resemble each other to a certain extent, but the young Lanner in which state it would be most likely to occur, may always be distinguished from the young Peregrine by its whiter head, and the adult has more brown on the head and neck.

The Peregrine is included in Professor Ansted's list, but only marked as occurring in Guernsey and Sark. There is no specimen at present in the Museum.

6. HOBBY. *Falco subbuteo*, Linnaeus. French, "Le Hobereau." The Hobby can only be considered as a rather rare occasional visitant, just touching the Islands on its southern migration in the autumn, and late in the autumn, for Mr. MacCulloch informs me that a Hobby was killed in the Islands, probably Guernsey, in November, 1873, and Mr. Couch, writing to me on the 10th of November, told me he had had a Hobby brought to him on the 8th of the same month. Both of these occurrences seem rather late, but probably the Hobby only touches the Islands for a very short time on passage, and quite towards the end of the migratory period. I do not know of any instance of the Hobby having occurred in the Islands on its northern migration in the spring, or of its remaining to breed.

It is included in Professor Ansted's list, and only marked as occurring in Guernsey. There is no specimen in the Museum.

7. MERLIN. *Falco aesalon*, [5] Bris., 1766. French, "Faucon Emirillon."—The pretty little Merlin is a much more common autumnal visitant to the Islands than the Hobby, but, like the Peregrine, the majority of instances are young birds of the year which visit the Islands on their autumnal migration. When I was in Guernsey in November, 1875, two Merlins, both young birds, were brought in to Mr. Couch's. Both were shot in the Vale, and I saw a third near Cobo, but did not shoot it. This also was a young bird. In some years Merlins appear to be more numerous than in others, and this seems

to have been one of the years in which they were most numerous. Unlike the Hobby, however, the Merlin does occasionally visit the Islands in the spring, as I saw one at Mr. Jago's, the bird-stuffer in Guernsey, which had been killed at Herm in the spring of 1876. This is now in the collection of Mr. Maxwell, the present owner of Herm. Though the Merlin visits the Islands both in the spring and autumn, I do not know that there is any instance of its having remained to breed, neither do I know of an occurrence during the winter. In the 'Zoologist' for 1875 Mr. Couch, in a communication dated November 29th, 1874, says—"A Merlin—a female—was shot in the Marais, which had struck down a Water Rail a minute or two before it was shot. After striking down the Rail the Merlin flew into a tree, about ten yards from which the man who shot it found the Rail dead. He brought me both birds. The skin of the Rail was broken from the shoulder to the back of the skull."

The more common prey, however, of the Merlin during the time it remains in the Islands is the Ring Dotterell, which at that time of year is to be found in large flocks mixed with Purres and Turnstones in all the low sandy or muddy bays in the Islands.

The Merlin is included in Professor Ansted's list, but only marked as occurring in Guernsey. There is no specimen in the Museum at present.

8. KESTREL. *Falco tinnunculus*, Linnaeus. French, "Faucon cresserelle."—The Kestrel is by far the commonest hawk in the Islands, and is resident throughout the year. I do not think that its numbers are at all increased during the migratory season. It breeds in the rocky parts of all the Islands. The Kestrel does not, however, show itself so frequently in the low parts—even in the autumn—as on the high cliffs, so probably Ring Dotterell, Purres, and Turnstones do not form so considerable a part of its food as they do of the Merlin. Skylarks, Rock and Meadow Pipits, and, in the summer, Wheatears, with a few rats and mice, seem to afford the principal food of the Kestrel, and to obtain these it has not to wander far from its breeding haunts.

The Kestrel is quite as common in Alderney and Herm, and even in the little Island of Jethou, as it is in Guernsey and Sark. One or two pairs, perhaps more, breed on the before-mentioned rock close to Jethou "La Fauconnihre," though a few pairs of Kestrels breeding there would scarcely have been sufficient to give it its name.

It is mentioned in Professor Ansted's list, but only marked as occurring in Guernsey and Sark. There are two specimens, a male and female, in the Museum.

9. SPARROWHAWK. *Accipiter nisus*, Linnaeus. French, "L'Epervier," "Tiercelet." — The Sparrowhawk, though a resident species and breeding in the Islands, is by no means so common as the Kestrel. In fact, it must certainly be considered rather a rare bird, which perhaps is not to be wondered at, as it is a more tree-breeding bird and less given to nesting amongst the rocks than the Kestrel. It does so sometimes, however, as I saw one fly out of some ivy-covered rocks near Petit Bo Bay the last time I was in the Islands on the 27th of May, 1878. I am certain this bird had a nest there, though the place was too inaccessible to be examined closely. The trees, however, at the Vallon or Woodlands would be much more likely nesting-places, especially as it might have an opportunity of appropriating a deserted nest of a Magpie or a Wood Pigeon, rather a favourite nesting-place of the Sparrowhawk.

Professor Ansted includes the Sparrowhawk in his list, but confines it to Guernsey and Sark; and probably, as a resident and breeding bird, he is right as far as my district is concerned, but I should think it must occasionally occur both in Alderney and Herm, though I have never seen a specimen from either Island, nor have I seen the bird about alive in either. There is one specimen in the Museum.

10. COMMON BUZZARD. *Buteo vulgaris*, Leach. French, "Buse." — The Buzzard is a tolerably regular, and by no means uncommon, autumnal visitant, specimens occurring from some of the

Islands almost every autumn. But it is, I believe, an autumnal visitant only, as I do not know of a single specimen taken at any other time of year, nor can I find a record of one. I have seen examples in the flesh from both Alderney and Herm, in both of which Islands it occurs at least as frequently as it does in Guernsey, though still only as an autumnal visitor.

It is included in Professor Ansted's list, but only marked as occurring in Guernsey, and there is one specimen in the Museum.

11. ROUGHLEGGED BUZZARD. *Buteo lagopus*, Gmelin. French, "Archibuse pattue" or "Buse pattue."—Though its visits seem not so absolutely confined to the autumn as the Common Buzzard, the Rough-legged Buzzard is a much more uncommon visitor to the Channel Islands, and can only be looked upon as a rare occasional straggler. Mr. MacCulloch informs me that one was killed near L'Hyvreuse, which is perhaps now more commonly known as the New Ground, in Guernsey, about Christmas, 1870, and I found one at the bird-stuffer and carpenter's shop at Alderney, which had been shot by his friend who shot the Greenland Falcon, but I could get no information about the date except that it was late autumn or winter, and about two years ago. These are the only Channel Island specimens of which I have been able to glean any intelligence. Probably, however, it has occurred at other times and been overlooked. As it may have occasionally been mistaken for the more common Common Buzzard, I may say that it is always to be distinguished from that bird by the feathered tarsus. On the wing, perhaps, when flying overhead, the most readily observed distinction is the dark band on the lower part of the breast. I have, however, seen a very dark variety of the Rough-legged Buzzard, in which nearly the whole of the plumage was a uniform dark chocolate-brown, and consequently the dark band on the breast could not be seen even when one had the bird in one's hand, and had it not been for the feathered tarsus this bird might easily have been mistaken for a very dark variety of the Common Buzzard, and when on the wing it would have been impossible to identify it. Indeed, though it was immediately distinguishable from the Common Buzzard by its feathered legs, there

was some little difficulty about identifying it, even when handling it as a skin.

Professor Ansted includes the Rough-legged Buzzard in his list, but only marks it as occurring in Guernsey. There is no specimen at present in the Museum.

12. MARSH HARRIER. *Circus æruginosus*, Linnaeus. French, "Busard des Marais."—This seems to be the least common of the Harriers in the Channel Islands, though it does occur occasionally, and perhaps more frequently than is generally supposed.

There are two specimens in the Museum in Guernsey both in immature plumage; in that state, in fact, in which this bird most commonly occurs, and in which it is the Bald Buzzard of Bewick.

Miss C.B. Carey records one in the November number of the 'Zoologist' for 1874 in the following words:—"In the May of this year an adult male Marsh Harrier was found in Herm. Unfortunately it got into the hands of some person who, I believe, kept it too long before bringing it over to be preserved, so that all that remains of it is the head." I had no opportunity of examining this bird myself, not even the head, but I am disposed to doubt its being fully adult, as it seems to me much more probable that it was much in the same state as those in the Museum, in which state it is much more common than in the fully adult plumage. Miss Carey seems only to have seen the head herself, so there may easily have been a mistake on this point.

Mr. MacCulloch writes me word that a Marsh Harrier was killed in Herm in May, 1875. It may be just possible, however, that this is the same bird recorded by Miss C.B. Carey, and that Mr. MacCulloch only heard of it in the May of the following year, and noted it accordingly. This, however, is mere supposition on my part, for which I have no reason except that both birds were said to have been killed in Herm, and both in May.

Professor Ansted mentions the Marsh Harrier in his list, but marks it as only found in Guernsey.

12. HEN HARRIER. *Circus cyaneus*, Linnaeus. French, "Busard St. Martin." [6] —The Hen Harrier, perhaps, occurs rather more frequently than the Marsh Harrier, but it can only be considered a rare occasional visitant. In June, 1876, I saw one young Hen Harrier, which had been shot in Herm in the April of that year, about the same time as the Iceland Falcon, and by the same keeper, who had brought it to Mr. Couch to stuff. Another was shot in Herm on the 19th of June, 1877. This bird is now in Mr. Maxwell's collection, where I saw it on the 27th of June. It was first reported to me by Mr. Jago, the bird-stuffer in Guernsey.

These are the only two Channel Island specimens of the Hen Harrier which I have been able to find. I have never shot it myself or seen it alive. It is, however, included in Professor Ansted's list, but marked as occurring in Guernsey only.

[13. Omitted.]

14. MONTAGU'S HARRIER. *Circus cineraceus*, Montagu. French, "Busard Montagu," "Busard cendri." —Montagu's Harrier is certainly a more frequent visitant to the Islands than either the Hen Harrier or the Marsh Harrier. Miss C.B. Carey records one in the 'Zoologist' for 1873 as having been shot in Alderney in July of that year. She adds that it was an adult male in full plumage, and that she saw it herself at Mr. Couch's shop. In the 'Zoologist' for 1874 she records another Montagu's Harrier—a young one—shot in Herm in July of that year. She adds that—"It was brought to Mr. Couch to skin. He found a whole Lark's egg, and also the shell of another, in its throat. He showed me how the whole egg was sticking in the empty shell of the broken one."

All the Harriers seem to have a special liking for eggs. In his notice of the Marsh Harrier Professor Newton says, in his edition of Yarrell,' that birds' eggs are an irresistible delicacy; and, in speaking of the food of the present species, he says it consists chiefly of grasshoppers, reptiles, small mammals, birds and their eggs; these last, if their size permit, being often swallowed whole, as was the case in the instance mentioned by Miss Carey. Mr. Howard Saunders also says he can bear witness to the egg-eating propensities of the Harriers.

Besides the two recorded by Miss C.B. Carey, I saw one—a young bird—in Mr. Maxwell's collection, which had been killed at Herm, and another—a young male—at Mr. Jago's, the bird-stuffer, which had also been killed at Herm. There were also two young birds in the bird-stuffer and carpenter's shop at Alderney, both of which had been killed in that Island shortly before my last visit, June, 1878.

As mistakes may occasionally arise in identifying specimens, especially in immature plumage, it may be as well to notice a distinction between the Hen Harrier and Montagu's Harrier, which has been pointed out by Mr. Howard Saunders, and which holds good in all ages and in both sexes. This distinction is, that in the Hen Harrier the outer web of the fifth primary is notched, whereas in Montagu's Harrier it is plain, or, in other words, the Hen Harrier has the exterior web of the primaries, up to and including the fifth, notched, and in Montagu's Harrier this is only the case as far as the fourth. [7] This distinction is very useful in identifying young birds and females, which are sometimes very much alike. In fully adult males the orange markings on the flanks and thighs, and the greyish upper tail-coverts of Montagu's Harrier, distinguish it immediately at a glance from the Hen Harrier, in which those parts are white.

Montagu's Harrier is not included by Professor Ansted in his list, nor is there a specimen in the Museum.

15. LONGEARED OWL. *Asiootus*, Linnaeus. French, "Hibou vulgaire," "Hibou moyen due."—The Long-eared Owl seems only a very rare and accidental visitor to the Channel Islands. I have never met with it myself, but Mr. Couch records the occurrence of one in the 'Zoologist' for 1875, p. 4296:—"I have a Long-eared Owl, shot at St. Martin's on the 9th of November in that year." This is the only occurrence I can be sure of, except that Mr. Couch, about two years

afterwards, sent me a skin of a Guernsey-killed Long-eared Owl; but this may have been the bird mentioned above, as he sent me no date with it.

As it is partially migratory, and its numbers in the British Islands, especially in the Eastern Counties, are increased during the autumn by migratory arrivals, a few may wander, especially in the autumn, to the Channel Islands, but it can only be rarely.

Professor Ansted includes it in his list, and marks it as having been found both in Guernsey and Sark. There is no specimen of the Long-eared Owl at present in the Museum. If there has been one it must have got moth-eaten, like many of the other birds there, and been destroyed.

16. SHORTEARED OWL. *Asio accipitrinus*, Pallas. French, "Hibou brachytte." — Unlike the Long-eared Owl, the Short-eared Owl is a regular autumnal visitant to the Channel Islands, arriving about October in considerable numbers, but remaining only for a short time, as I do not know of any making their appearance after the end of November, and the majority of those that have arrived seem to pass on about that time, not remaining throughout the winter, and I hear of no instances of their occurring on the spring migration, so the majority must pass north by a different line from that pursued by them on the southern migration.

There is only one specimen at present in the Museum. Professor Ansted mentions it in his list, but only as found in Guernsey and Sark; but it is quite as common in Alderney, from which Island I have seen specimens, and I think also from Herm, but I cannot be quite sure about this, though of course there can be no reason why it should not be found there, as Herm is only three miles as the crow flies from Guernsey.

17. BARN OWL. *Aluco flammeus*, Linnaeus. French, "Chouette effraie." — I have never seen the Barn or Yellow Owl alive in the Channel Islands myself, but Mr. MacCulloch does not consider it at all

rare in Guernsey, and Mr. Jago informs me the Barn Owls have taken possession of a pigeon-hole in a house in the Brock Road opposite his, and that he sees and hears them every night. Some years ago he told me he shot one near the Queen's Tower. He was not scared like the man who shot one in the churchyard, and thought he had shot a cherubim, but he had to give up shooting owls, as the owner of the pigeon-hole where the owls have taken up their abode remonstrated with him, and he has since refrained, though he has had several chances. The vacancy caused by the one being shot was soon filled up.

The Barn Owl is mentioned in Professor Ansted's list, and restricted to Guernsey and Sark. There are two specimens in the Museum, both of which are said to have been killed in Guernsey.

18. REDBACKED SHRIKE. *Lanius Collurio*, Linnaeus. French, "Pie-grieche icorcheur." — The Red-backed Shrike may be considered a tolerably regular, but not very common, summer visitor to the Channel Islands. In June, 1876, I several times saw a male bird about the Vallon, in Guernsey. The female no doubt had a nest at the time in the Vallon grounds, but I could not then get in there to search for it.

As the Red-backed Shrike frequently returns to the same place every year, I expected again to find this bird, and perhaps the female and the nest this year, 1878, about the Vallon, but I could see nothing of either birds or nest, though I searched both inside and outside the Vallon grounds.

Young Mr. Le Cheminant, who lives at Le Ree and has a small collection of Guernsey eggs mostly collected by himself in the Island, had one Red-backed Shrike's egg of the variety which has the reddish, or rather perhaps pink, tinge. There were also some eggs in a Guernsey collection in the Museum. These were all of the more ordinary variety. There were also two skins — a male and female — in the Museum. The bird seems rather local in its distribution about the Island, as I never saw one about the Vale in any of my visits, not even this year, 1878, when I was there for two months, and had ample opportunity of observing it had it been there. There are, however, plenty of places nearly as well suited to it in the Vale as about the Vallon or Le Ree. I have never seen it in either of the other Islands, though no doubt it occasionally occurs both in Sark and Herm, if not in Alderney.

Professor Ansted includes the Red-backed Shrike in his list, and marks it only as occurring in Guernsey. I have no evidence of any other Shrike occurring in the Islands, though I should think the Great Grey Shrike, *Lanius excubitor*, might be an occasional autumn or winter visitor to the Islands; but I have never seen a specimen myself or been able to glean any satisfactory information as to the occurrence of one, either from the local bird-stuffers or from Mr. MacCulloch, or any of my friends who have so kindly supplied me with notes; neither does Professor Ansted mention it in his list.

19. SPOTTED FLYCATCHER. *Muscicapa grisola*, Linnaeus. French, "Gobe-mouche gris."—The Spotted Flycatcher is a regular and numerous summer visitor, generally quite as numerous in certain localities as in England, its arrival and departure being about the same time. It occurs also in Sark and Herm, and probably in Alderney, but I do not remember having seen one there. In Guernsey it is perhaps a little local in its distribution, avoiding to a great extent such places as the Vale and the open ground on the cliffs, but in all the gardens and orchards it is very common.

Spotted Flycatchers appear, however, to vary in numbers to a certain extent in different years. This year, 1878, they came out in great force, especially on the lawn at Candie where they availed themselves to a large extent of the croquet-hoops, from which they kept a good look-out either for insects on the wing or on the ground, and they might be as frequently seen dropping to the ground for some unfortunate creeping thing that attracted their attention as rising in the air to give chase to something on the wing. Certainly, when I was in Guernsey about the same time in 1866, Spotted Flycatchers did not appear to be quite so numerous as in 1878. This was probably only owing to one of those accidents of wind and weather which render migratory birds generally, less numerous in some years than they are in others, however much they may wish and endeavour, which seems to be their usual rule, to return to their former breeding stations.

Professor Ansted mentions the Spotted Flycatcher in his list, but does not add, as he usually does, any letter showing its distribution through the Islands. This probably is because it is generally distributed through them all. There is no specimen in the Museum.

20. GOLDEN ORIOLE. *Oriolus galbula*, Linnaeus. French, "Le Loriot."—I have never seen the bird alive or found any record of the occurrence of the Golden Oriole in Guernsey or the neighbouring Islands, and beyond the fact that there was one example—a female—in the Museum (which may have been from Jersey) I had been able to gain no information on the subject except of a negative sort. No specimen had passed through the hands of the local bird-stuffers certainly for a good many years, for Mr. Jago's mother who about twenty or thirty years ago, when she was Miss Cumber, had been for some considerable time the only bird-stuffer in the Island, told me she did not know the bird, and had never had one through her hands. It seemed to me rather odd that a bird which occurs almost every year in the British Islands, occasionally even as far west as Ireland, as a straggler, and which is generally distributed over the continent of Europe in the summer, should be totally unknown in the Channel Islands. Consequently writing to the 'Star' about another Guernsey bird—a Hoopoe—which had been recorded in that paper, I asked for information as to the occurrence of the Golden Oriole in the Islands, and shortly after the following letter signed "Tereus" [8] appeared in the 'Star':—"Concerning the occurrence of the Golden Oriole I cannot speak from my own personal knowledge, but I believe there can be no doubt that the bird has been occasionally seen here. Its presence, however, must be much more rare than that of the Hoopoe, for a bird of such plumage as the Oriole would be more likely to attract even more attention than the comparatively sober-coloured Hoopoe, and if half so common as the latter would be sure to fall before the gun of the fowler. There was a specimen of the female bird in the Museum of the Mechanics' Institution, but I am not sure about its history, and I have some

reason to suppose it was shot in Jersey. Our venerable national poet, Mr. George Mitivier, has many allusions to the Oriole in his early effusions, whether written in English, French, or our vernacular dialect. It seems to have been an occasional visitor at St. George's; but in Mr. Mitivier's early days the island was far more wooded than it is at present, and it is possible that the wholesale destruction of hedgerow elms and the grubbing-up of so many orchards in order to employ the ground more profitably in the culture of early potatoes and brocoli, by which the island has lost much of its picturesque beauty, may have had the effect of deterring some of the occasional visitors from alighting here in their periodical migrations." Signed "Tereus."

A short time after the appearance of this letter in the 'Star' on the 16th of May, 1878, Mr. MacCulloch himself wrote to me on the subject and said: — "I had yesterday a very satisfactory interview with Mr. George Mitivier. He is now in his 88th or 89th year. He told me he was about thirteen when he went to reside with his relations, the Guilles, at St. George. There was then a great deal of old timber about the place and a long avenue of oaks, besides three large cherry orchards. One day he was startled by the sight of a male Oriole. He had never seen the bird before. Whether it was that one that was killed or another in a subsequent year I don't know, but he declares that for several years afterwards they were seen in the oak trees and among the cherries, and that he has not the least doubt but that they bred there. One day an old French gentleman of the name of De l'Huiller from the South of France, an emigrant, noticed the birds and made the remark — 'Ah! vous avez des loriots ici; nous en avons beaucoup chez nous, ils sont grands gobeurs de cerises.' It would appear from this that cherries are a favourite food with this bird, and the presence of cherry orchards would account for their settling down at St. George. I believe they are said to be very shy, and the absence of wood would account for their not being seen in the present day."

I have no doubt that Mr. MacCulloch is right that the cherry orchards, to say nothing of other fruit trees, tempted the Golden Orioles to remain to breed in the Island, for they are "grand gobeurs" not only of "cerises," but of many other sorts of fruit, particularly of grapes and figs — in grape countries, indeed, doing a deal of damage amongst the vineyards. This damage to grapes would not, however, be much felt in Guernsey, as all the grapes are protected by orchard-houses. But though the grapes are protected, and most, if not all, the cherry orchards cut down, still there is plenty of unprotected fruit in Guernsey to tempt the Golden Oriole to remain in the Islands, and to bring the wrath and the gun of the gardener both to bear upon him when he is there. This, however, only shows that from the time spoken of by Mr. Mitivier down to the present time very few Golden Orioles could have visited Guernsey, and still fewer remained to breed; for what with their fruit-eating propensities and their bright plumage, hardly a bird could have escaped being shot and subsequently making its appearance in the bird-stuffers' windows, and affording a subject for a notice in the 'Star,' or some other paper. I think therefore, on the whole, that though Guernsey still affords many temptations to the Golden Oriole, and is sufficiently well-wooded to afford shelter to suit its shy and suspicious habits, yet for some reason or other the bird has not visited the Island of late years even as an accidental visitant, or, if so, very rarely.

The Golden Oriole is mentioned in Professor Ansted's list, and marked as having occurred in Guernsey and Sark, but nothing more is said about the bird. Probably Guernsey was mentioned as a locality on account of the female specimen in the Museum, but with this exception I have never heard of its making its appearance in Sark even as a straggler.

21. DIPPER. *Cinclus aquaticus*, Bechstein. French, "Aquassihre," "Cincle plongeur."—The Dipper or Water Ouzel, though not very common, less so, indeed, than the Kingfisher, is nevertheless a resident species, finding food all through the year in the clear pools left by the tide, and also frequenting the few inland ponds, especially the rather large ones, belonging to Mr. De Putron in the Vale, where there is always a Dipper or a Kingfisher to be seen, though I do not think the Dipper ever breeds about those ponds—in fact there is no place there which would suit it; but though I have never found the nest myself in Guernsey, I have been informed, especially by Mr. Gallienne, that the Dipper makes use of some of the rocky bays,

forming his nest amongst the rocks as it would on the streams of Dartmoor and Exmoor.

Captain Hubboch, however, writes me word he saw one in Alderney in the winter of 1861-62, and there seems no reason why a few should not remain there throughout the year as in Guernsey.

All the Guernsey Dippers I have seen, including the two in the Museum, which are probably Guernsey-killed, have been the common form, *Cinclus aquations*. The dark-breasted form, *Cinclus melanogaster*, may occur as an occasional wanderer, though the Channel Islands are somewhat out of its usual range. There being no trout or salmon to be protected in Guernsey, the Dipper has not to dread the persecution of wretched keepers who falsely imagine that it must live entirely by the destruction of salmon and trout ova, though the contrary has been proved over and over again.

Professor Ansted includes the Dipper in his list, but only marks it as occurring in Guernsey.

22. MISTLETOE THRUSH. *Turdus viscivorus*, Linnaeus. French, "Merle Draine," "Grive Draine."—I quite agree with the remarks made by Professor Newton, in his edition of 'Yarrell,' as to the proper English name of the present species, and that it ought to be called the Mistletoe Thrush. I am afraid, however, that the shorter appellation of Missel Thrush will stick to this bird in spite of all attempts to the contrary. In Guernsey the local name of the Mistletoe Thrush is "Geai," by which name Mr. Mitivier mentions it in his 'Dictionary of Guernsey and Norman French.' He also adds that the Jay does not exist in this Island. This is to a certain extent confirmed by Mr. MacCulloch, who says he is very doubtful as to the occurrence of the Jay in the Island, and adds that the local name for the Mistletoe Thrush is "Geai." Mr. Gallienne, in a note to Professor Ansted's list, confirms the scarcity of the Jay, as he says the Rook and the Jay are rarely seen here, although they are indigenous to Jersey. The local name "Geai" may perhaps have misled him as to the occasional appearance of the Jay. I have never seen a real Jay in Guernsey myself.

As far as I am able to judge from occasional visits to the Island for the last thirty years the Mistletoe Thrush has greatly increased in numbers in Guernsey, especially within the last few years, and Mr. MacCulloch and others who are resident in the Island quite agree with me in this. I do not think its numbers are much increased at any time of year by migrants, though a few foreigners may arrive in the autumn, at which time of year considerable numbers of Mistletoe Thrushes are brought into the Guernsey market, where they may be seen hanging in bunches with Common Thrushes, Redwings, Blackbirds, Fieldfares, Starlings, and an occasional Ring Ouzel. Fieldfares and Mistletoe Thrushes usually sell at fourpence each, the rest at fourpence a couple.

Professor Ansted mentions it in his list, but confines it to Guernsey and Sark. This is certainly not now the case, as I have seen it nearly as numerous in Alderney and Herm as any of the other Islands. There is a specimen in the Museum.

23. SONG THRUSH. *Turdus musicus*, Linnaeus. French, "Grive," "Merle Grive."—Very common and resident in all the Islands, and great is the destruction of snails by Thrushes and Blackbirds—in fact, nowhere have I seen such destruction as in the Channel Islands, especially in Guernsey and Herm, where every available stone seems made use of, and to considerable purpose, to judge from the number of snail-shells to be found about; and yet the gardeners complain quite as much of damage to their gardens, especially in the fruit season, by Blackbirds and Thrushes, as the English gardeners and seem equally unready to give these birds any credit for the immense destruction of snails, which, if left alone, would scarcely have left a green thing in the garden.

The local name of the Thrush is "Mauvis." It is, of course, included in Professor Ansted's list, but with the Fieldfare, Redwing, and Blackbird, marked as only occurring in Guernsey and Sark. All these birds, however, are equally common in Alderney, Herm, and Jethou. There is also a specimen of each in the Museum.

24. REDWING. *Turdus iliacus*, Linnaeus. French, "Grive mauvis," "Merle mauvis."—A regular and numerous winter visitant to all the Islands, arriving about the end of October, and those that are not shot and brought into the market departing again in March and April.

25. FIELDFARE. *Turdus pilaris*, Linnaeus. French, "Grive litorne," "Merle litorne."—Like the Redwing, the Fieldfare is a regular and numerous winter visitant, and arrives and departs about the same time.

When in Guernsey in November, 1871, I did not see either Redwings or Fieldfares till a few days after my arrival on the 1st; after that both species were numerous, and a few days later plenty of them might be seen hanging up in the market with the Thrushes and Blackbirds, but for the first few days there were none to be seen there. Probably this was rather a late year, as neither bird could have arrived in any numbers till the first week in November, and in all probability not till towards the end of the week.

26. BLACKBIRD. *Turdus merula*, Linnaeus. French, "Merle noir."—
- The Blackbird is a common and numerous resident in all the Islands in the Bailiwick of Guernsey. The Guernsey gardeners, like their brethren in England, make a great fuss about the mischief done by Blackbirds in the gardens, and no doubt Blackbirds, like the Golden Orioles, are "grand gobeurs" of many kinds of fruit; but the gardeners should remember that they are equally "grand gobeurs" of many kinds of insects as well, many of the most mischievous insects to the garden, including wasps (I have myself several times found wasps in the stomach of the blackbird) forming a considerable portion of their food, the young also being almost entirely fed upon worms, caterpillars, and grubs; and when we remember that it is only for a short time of the year that the Blackbird can feed on

fruit, which in most cases can be protected by a little care, and that during the whole of the other portion of the year it feeds on insects which would do more damage in the garden than itself, it will be apparent that the gardener has really no substantial ground of complaint.

As in England, variations in the plumage of the Blackbird are not uncommon. I have one Guernsey specimen of a uniform fawn colour, and another rather curiously marked with grey, the tail-feathers being striped across grey and black. This is a young bird recently out of the nest, and I have no doubt would, after a moult or two, have come to its proper plumage, probably after the first moult, as seems to me frequently the case with varieties of this sort, though I have known a Blackbird show a good deal af white year after year in the winter, resuming its proper plumage in the summer; and Mr. Jago mentions a case of a Blackbird which passed through his hands which was much marked with grey. This bird was found dead, and the owner of the estate on which it was found informed Mr. Jago that it had frequented his place for four years, and that he had seen it with its mate during the summer; so in this case the variation certainly seems to have been permanent.

27. RING OUZEL. *Turdus torquatus*, Linnaeus. French, "Merle ` plastron."—I do not think the Ring Ouzel is ever as common in the Channel Islands as it is on migration in South Devon. A few, however, make their appearance in each of the Islands every autumn, but they are never very numerous, and do not remain very long, arriving generally about the end of September and remaining till the end of November or beginning of December, during which time a few may always be seen hung up in the market. Many of the autumnal arrivals are young birds of the year, with the white crescent on the breast nearly wanting or only very faintly marked.

Mr. Gallienne, in his remarks appended to Professor Ansted's list, says the Ring Ouzel stays with us throughout the year, but is more plentiful in winter than in summer. But I have never myself seen one either dead or alive in the spring or summer. It may, however, occasionally visit the Island in the spring migration, but I know of no authentic instance of its remaining to breed, nor have I seen the eggs in any Guernsey collection. I have seen specimens of the Ring Ouzel from Alderney, and it appears to me about equally common at the same time of year in all the Islands. Mr. MacCulloch, however, writes to me:—"From what I have heard the Ring Ouzel is more common in Alderney than Guernsey, where it is seen mostly on the southern cliffs." The south end of the Island is no doubt its favourite resort in Guernsey. As far as Alderney is concerned Captain Hubback, R.A., who has been quartered there at different times, says he has never seen one there; but I do not think he has been much there in the early autumn.

Professor Ansted includes it in his list, and marks it as occurring in Guernsey and Sark. There are several, both male and female and young, in the Guernsey Museum.

28. HBDGESPARROW. *Accentor modularis*, Linnaeus. French, "Mouchet," "Tranne buisson," "Accenteur mouchet."—The Hedgesparrow is, I think, quite as common as in England, and resident throughout the year in all the Islands. According to Mr. Mitivier's 'Dictionary' its local name is "Verdeleu," and he describes it as "Oiseau qui couvre les oeufs de Coucou." In Guernsey, however, Cuckoos are much too numerous for the Hedgesparrow to afford accommodation for them all.

Professor Ansted mentions the Hedgesparrow in his list, but restricts it to Guernsey and Sark. I have, however, frequently seen it in Alderney and Herm, and the little Island of Jethou.

29. ROBIN. *Ericathus rubecula*, Linnaeus. French. "Bec-fin rougegorge," "Rouge gorge." The Robin, like the Hedgesparrow, is a common resident in all the Islands, and I cannot find that its numbers are increased at any time of year by migration. But on the other hand I should think a good many of the young must be driven off to seek quarters elsewhere by their most pugnacious parents, for of all birds the Robin is by far the most pugnacious with which I am ac-

quainted, and deserves the name of "pugnax" much more than the Ruff, and in a limited space like Jethou and Herm battles between the old and the young would be constant unless some of the young departed altogether from the Island.

Professor Ansted includes the Robin in his list, but, as with the Hedgesparrow, only mentions it as occurring in Guernsey and Sark. It is, however, equally common in Alderney, Jethou, and Herm.

30. REDSTART. *Ruticilla phoenicurus*, Linnaeus. French, "Rouge-queue," "Bec-fin des murailles."—I should not have included the Redstart in this list, as I have never seen it in the Islands myself, but on sending a list of the birds I intended to include to Mr. MacCulloch, he wrote to say—"You mention Tithy's Redstart; the common one is also seen here." In consequence of this information I looked very sharply out for the birds during the two months (June and July) which I was in Guernsey this year (1878), but I never once saw the bird in any of the Islands, nor could I find any one who had; and such a conspicuous and generally well known bird could hardly have escaped observation had it been in the Island in any numbers. I may add that I have had the same bad luck in all my former visits to the Islands, and never seen a Redstart. I suppose, however, from Mr. MacCulloch's note that it occasionally visits the Islands for a short time on migration, very few, if any, remaining to breed.

It is included in Professor Ansted's list, but only marked as occurring in Guernsey. There is, however, no specimen at present in the Museum.

31. BLACK REDSTART. *Ruticilla titys*, Scopoli. French, "Rouge queue Tithys."—The Black, or Tithys Redstart, as it is sometimes called, is a regular and by no means uncommon autumnal visitant to Guernsey. It seems very much to take the place of the Wheatear, arriving about the time the Wheatear departs, and mostly frequenting the same places. In Guernsey it is most common near the sea about the low part of the Island, from L'ancresse Common to Per-

relle Bay. In habits it puts one very much in mind of the Wheatear, being very fond, like that bird, of selecting some big stone or some other conspicuous place to perch on and keep a look-out either for intruders or for some passing insect, either flying or creeping, for it is an entirely insect-feeding bird.

I have never seen the Black Redstart about the high part of the Island amongst the rocks, which I am rather surprised at, as in the south coast of Devon it seems particularly partial to high cliffs and rocks, such as the Parson and Clerk Rock near Teignmouth; but in Guernsey the wild grassy commons, with scattered rocks and large boulders, and occasionally a rough pebbly beach, especially the upper part of it where the pebbles join the grass, seem more the favourite resort of this bird than the high rocks, such places probably being more productive of food. It is of course quite useless to look for this bird in the interior of the Island in gardens and orchards, and such places as one would naturally look for the Common Redstart.

The male Black Redstart may be immediately distinguished from the Common Redstart by the black breast and belly, and by the absence of the white mark on the forehead. The male Black Redstart has also a white patch on the wing caused by the pale, nearly white, margins of the feathers. The females are more alike, but still may easily be distinguished, the general colour of the female Black Redstart being much duller — a dull smoke-brown instead of the reddish brown of the Common Redstart.

Some slight variations of plumage take place in the Black Redstart at different ages and seasons, which have led to some little difficulties, and to another supposed species, *Ruticilla cairii* of Gerbe being suggested, but apparently quite without reason. I have never seen the Black Redstart in the Islands at any time of year except the autumn, and do not know of its occurrence at any other time.

Professor Ansted includes it in his list, but gives no locality; and there is no specimen in the Museum.

32. STONECHAT. *Pratincola rubicola*, Linnaeus. French, "Tarier rubicole," "Traquet pbtre," "Traquet rubicole."—The Stonechat is a numerous and regular summer visitant, breeding in all the Islands, but I do not think any remain throughout the winter; of course a few scattered birds may occasionally do so in some sheltered locality, but I have never seen one in the Islands as late as November. Both in the Vale and on the Cliffs in the higher part of the Island the Stonechat is very common, and the gay little bird, with its bright plumage and sprightly manner, may be seen on the top of every furze bush, or on a conspicuous twig in a hedge in the wilder parts of the Island, but is not so common in the inland and more cultivated parts, being less frequently seen on the hedges by the roadside than it is here, Somersetshire, or in many counties in England. In Alderney it is quite as common as in Guernsey, and I saw two nests this year (1878) amongst the long grass growing on the earthworks near the Artillery Barracks; it is equally common also both in Jethou, Sark, and Herm.

There were a great many Stonechats in the Vale when I was there this year (1878). Generally they seemed earlier in their breeding proceedings than either Wheatears, Tree Pipits, or Sky Larks, which were the three other most numerous birds about that part of the Island, as there were several young ones about when we first went to live in the Vale early in June; still occasionally nests with eggs more or less hard sat might be found, but the greater number were hatched when fresh eggs of Tree Pipits and Sky Larks were by no means uncommon.

Professor Ansted includes the Stonechat in his list, but marks it as confined to Guernsey and Sark. There is a specimen in the Museum.

33. WHINCHAT. *Pratincola rubetra*, Linnaeus. French, "Tarier ordinaire," "Traquet tarier."—The Whinchat seems to me never so numerous as the Stonechat, and more local in its distribution during the time it is in the Islands. It is only a summer visitant, and I doubt if it always remains to breed, though it certainly does so occasionally, as I have seen it in Guernsey through June and July mostly in the south part of the Island, near Pleimont. In my last visit to the Islands, however, in June and July, 1878, I did not see the Whinchat anywhere, neither did I see one when there in June, 1876.

Professor Ansted includes the Whinchat in his list, and marks it as occurring in Guernsey and Sark. There is no specimen in the Museum.

34. WHEATEAR. *Saxicola Oenanthe,* Linnaeus. French, "Motteux cul blanc," "Traquet moteux."—A very common summer visitant to all the Islands, arriving in March and departing again in October, none remaining through the winter—at least, I have never seen a Wheatear in the Islands as late as November on any occasion. In the Vale, where a great many breed, the young began to make their appearance out of the nest and flying about, but still fed by their parents, about the 16th of June. In Guernsey it is rather locally distributed, being common all round the coast, both on the high and low part of the Island, but only making its appearance in the cultivated part in the interior as an occasional straggler. It is quite as common in Alderney and the other Islands as it is in Guernsey, in Alderney there being few or no enclosures, and no hedgerow timber. It is more universally distributed over the whole Island, in the cultivated as well as the wild parts.

Professor Ansted includes it in his list, but marks it as only occurring in Guernsey and Sark. There are several specimens in the Museum, but I did not see any eggs either there or in young Le Cheminant's collection. This is probably because in Guernsey the Wheatear has a great partiality for laying its eggs under large slabs and boulders of granite perfectly immovable; the stones forming one of the Druids' altars in the Vale, were made use of to cover a nest when I was there.

35. REED WARBLER. *Acrocephalus streperus,* Vieillot. French, "Rousserolle effarvatte," "Bec-fin des roseaux."—I did not find out the Reed Warbler as a Guernsey bird till this year (1878), though it is a rather numerous but very local summer visitant. But Mr. MacCulloch put me on the right track, as he wrote to me to say—"The Reed Warbler builds in the Grand Mare. I have seen several of their curi-

ous hanging nests brought from there." This put me on the right scent, and I went to the place as soon as I could, and found parts of it a regular paradise for Reed Warblers, and there were a considerable number there, who seemed to enjoy the place thoroughly, climbing to the tops of the long reeds and singing, then flying up after some passing insect, or dropping like a stone to the bottom of the reed-bed if disturbed or frightened. On my first visit to the Grand Mare I had not time to search the reed-beds for nests. But on going there a second time, on June 17, with Colonel l'Estrange, we had a good search for nests, and soon found one with four eggs in it which were quite fresh. This nest was about three feet from the ground, tied on to four reeds, [9] and, as usual, having no support at the bottom, was made entirely of long dry bents of rather coarse grass, and a little of the fluff of the cotton plant woven amongst the bents outside, but none inside. We did not find any other nests in the Grand Mare, though we saw a great many more birds; the reeds, however, were very thick and tall, high over our heads, so that when we were a few feet apart we could not see each other, and the place was full of pitfalls with deep water in them, which were very difficult to be seen and avoided. Many of the nests, I suspect, were amongst the reeds which were growing out of the water. Subsequently, on July the 12th, I found another Reed Warbler's nest amongst some reeds growing by Mr. De Putron's pond near the Vale Church; this nest, which was attached to reeds of the same kind as those at the Grand Mare, growing out of water about a foot deep: it was about the same height above the water that the other was from the ground; it had five eggs in it hard sat. There were one or two pairs more breeding amongst these reeds, though I could not very well get at the place without a boat, but the birds were very noisy and vociferous whenever I got near their nests, as were the pair whose nest I found. There were also a few pairs in some reed-beds of the same sort near L'Eree.

These are all the places in which I have been able to find the Reed Warbler in Guernsey. I have not found it myself in Alderney, but Mr. Gallienne, in his remarks published with Professor Ansted's list, says: — "I have put the Reed Wren as doubtful for Guernsey, but I have seen the nest of this bird found at Alderney." In the list itself it is marked as belonging to Guernsey, Alderney, and Sark.

The Reed Warbler, though entirely insectivorous, is a very tame and amusing cage-bird, and may easily be fed on raw meat chopped fine and a little hard-boiled egg; but its favourite food is flies, and of these it will eat any quantity, and woe even to the biggest bluebottle that may buzz through its cage, for the active little bird will have it in a moment, and after a few sharp snaps of the beak there is quite an end of the bluebottle. Daddy long-legs, too, are favourite morsels, and after a little beating about disappear down the bird's throat—legs, wings, and all, without any difficulty. The indigestible parts are afterwards cast up in pellets in the same manner as with Hawks.

I have never seen the nearly-allied and very similar Marsh Warbler, *Acrocephalus palustris*, in Guernsey, but, as it may occasionally occur, it may be as well perhaps to point out what little distinction there is between the species. This seems to me to consist chiefly in the difference of colour, the Reed Warbler, *Acrocephalus streperus*, at all ages and in all states of plumage, being a warmer, redder brown than *Acrocephalus palustris*, which is always more or less tinged with green. The legs in *A. streperus* are always darker than in *A. palustris*; the beak also in *A. palustris* seems rather broader at the base and thicker. This bird also has a whitish streak over the eye, which seems wanting in *A. streperus*. These distinctions seem to me always to hold, good even in specimens which have been kept some time and have faded to what has now generally got the name of "Museum colour."

Mr. Dresser, in his 'Birds of Europe,' points out another distinction which no doubt is a good one in adult birds with their quills fully grown, but fails in young birds and in adults soon after the moult, before the quills are fully grown, and also before the moult if any quills have been shed and not replaced. This distinction is that in *A. streperus* the second (that is the first long quill, for the first in both species is merely rudimentary) is shorter than the fourth, and in *A. palustris* it is longer.

Though I think it not at all improbable that the Marsh Warbler, *Acrocephalus palustris*, may occur in Guernsey, I should not expect to find it so much in the wet reed-beds in the Grand Mare and at the Vale pond as amongst the lilac bushes and ornamental shrubs in the gardens, or in thick bramble bushes in hedgerows and places of that sort.

36. SEDGE WARBLER. *Acrocephalus schoenobaenus*, Linnaeus. French, "Bec-fin phragmite."—The Sedge Warbler is by no means so common as the Reed Warbler, though, like it, it is a summer visitant, and is quite as local. I did not see any amongst the reeds which the Reed Warbler delighted in, but I saw a few amongst some thick willow hedges with thick grass and rushes growing by the side of the bank, and a small running stream in each ditch. Though perfectly certain the birds were breeding near, we could not find the nests. So well were they hidden amongst the thick grass and herbage by the side of the stream that Colonel l'Estrange and myself were quite beaten in our search for the nest, though we saw the birds several times quite near enough to be certain of their identity. I did not shoot one for the purpose of identification, as perhaps I ought to have done, but I thought if I shot one it would be extremely doubtful whether I should ever find it amongst the thick tangle—certainly unless quite dead there would not have been a chance. I felt quite certain, however, that all I saw were Sedge Warblers; had I felt any doubt as to the possibility of one of them turning out to be the Aquatic Warbler, *Acrocephalus aquaticus*, I should certainly have tried the effect of a shot. As it is quite possible, however, that the Aquatic Warbler may occasionally, or perhaps regularly, in small numbers, visit the Channel Islands, as they are quite within its geographical range, I may point out, for the benefit of any one into whose hands it may fall, that it may easily be distinguished from the Sedge Warbler by the pale streak passing through the centre of the dark crown of the head.

The Sedge Warbler is not mentioned by Professor Ansted in his list, and there is no specimen of either this or the Reed Warbler in the Museum.

37. DARTFORD WARBLER. *Melizophilus undatus,* Boddaert. French, "Pitchou Provencal," "Bee-fin Pittechou."—The Dartford Warbler is by no means common in the Channel Islands—indeed I have never seen one there myself, but Miss C.B. Carey records one in the 'Zoologist' for 1874 as having been knocked down with a stone in the April of that year and brought into Couch's shop, where she saw it. I have no doubt of the correctness of this identification, as Miss Carey knew the bird well. I see no reason why it should not be more common in Guernsey than is usually supposed, as there are many places well suited to it, but its rather dull plumage, and its habit of hiding itself in thick furze-bushes, and creeping from one to another as soon as disturbed, contribute to keep it much out of sight, unless one knows and can imitate its call-note, in which case the male bird will soon answer and flutter up to the topmost twig of the furze-bush in which it may have previously been concealed, fluttering its wings, and repeating the call until again disturbed. This is the only occurrence of which I am aware in any of the Islands, included in the limits I have prescribed for myself; but Mr. Harvie Brown has recorded two seen by him near Grhve de Lecq, in Jersey, in January. See 'Zoologist' for 1869, p. 1561.

It is not included in Professor Ansted's list, and there is no specimen in the Museum.

38. WHITETHROAT. *Sylvia rufa,* Boddaert. French, "Fauvette grise," "Bec-fin Grisette."—The Whitethroat has hitherto perhaps been better known by the name used in the former edition of 'Yarrell' and by Messrs. Degland and Gerbe, *Curruca cinerea,* but in consequence of the inexorable rule of the British Association the name "*rufa,*" given by Boddaert in 1783, has now been accepted for this bird. I have not generally thought it necessary to point out these changes, but in this instance it seemed necessary to do so, as in the former edition of 'Yarrell' the Chiffchaff was called by the name *Sylvia rufa,* and this might possibly have caused some confusion unless the change had been pointed out.

The Whitethroat is by no means so common in the Channel Islands as it is in England, and though a regular summer visitant it only makes its appearance in small numbers. A few, however, may be seen about the fields and hedgerows in the more cultivated parts of the country. It certainly has not got the reputation for mischief in the garden it has in England, as none of the gardeners I asked about it, and who were complaining grievously of the mischief done by birds, ever mentioned the Whitethroat, or knew the bird when asked about it.

Professor Ansted includes the bird in his list, and restricts it to Guernsey, but I see no reason why it should not occur equally in Sark and Herm. There is no specimen at present in the Museum.

39. LESSER WHITETHROAT. *Sylvia curruca*, Linnaeus. French, "Bee-fin babillard."—Like the Whitethroat, the Lesser Whitethroat is a regular, but by no means a numerous summer visitant to Guernsey. I saw a few in the willow-hedges about the Grand Mare, and in one or two other places near there, and young Le Cheminant had one or two eggs in his collection, probably taken about L'Eree.

The Lesser Whitethroat is included in Professor Ansted's list, and only marked as occurring in Guernsey. There is at present no specimen in the Museum.

40. BLACKCAP. *Sylvia atricapilla*, Linnaeus. French, "Fauvette ` tjte noire," "Bec-fin ` tjte noire."—Though generally known as the Guernsey Nightingale, the Blackcap, though a regular, is by no means a numerous summer visitant. I have, however, always seen a few about every time I have been in the Island in the summer. There are a few eggs in the Museum, and in Le Cheminant's collection.

The Blackcap is mentioned by Professor Ansted in his list, and restricted to Guernsey. There is only one specimen—a female—at present in the Museum.

41. WILLOW WREN. *Phylloscopus trochilus*, Linnaeus. French, "Bee-fin Pouillat."—The Willow Wren is a tolerably numerous summer visitant, I believe, to all the Islands, though I have only seen it myself in Guernsey and Sark. In Guernsey I have seen it about the Grand Mare, and in some trees near the road about St. George, and about the Vallon on the other side of the Island. It remains all the summer and breeds.

Professor Ansted has not included it in his list, although it seems tolerably well known, and has a local name "D'moubiselle," which Mr. Mitivier, in his 'Dictionary,' applies to the Willow Wren of the English. This name, however, is probably equally applicable to the Chiffchaff.

42. CHIFFCHAFF. *Phylloscopus collybita*, Vieillot. French, "Bee-fin veloce."—The Chiffchaff is certainly more common in Guernsey than the Willow Wren. In Guernsey I have seen it in several places; about Candie, where a pair had a nest this summer in the mowing-grass before the house; near the Vallon; and about St. George. I have also seen it in Sark, but not in either of the other Islands, though no doubt it occurs in Herm, if not in Alderney.

It is mentioned by Professor Ansted as occurring in Guernsey and Sark. I have never seen the Wood Wren in Guernsey, and, judging from its favourite habitations here in Somerset, I should not think it at all likely to remain in the Channel Islands through the summer, though an occasional straggler may touch the Islands on migration. There is no specimen of either the Chiffchaff or Willow Wren in the Museum.

43. GOLDEN-CRESTED WREN. *Regulus cristatus*, Koch. French, "Roitelet ordinaire."—The Golden-crest is resident in the Islands, but not very numerous, and I doubt if its numbers are regularly increased in the autumn by migrants, as is the case in the Eastern Counties of England. Migratory flocks, however, sometimes make their appearance; and Mr. MacCulloch writes to me—"The Golden-

crest occasionally comes over in large flocks, apparently from Normandy, flying before bad weather. This, however, cannot be said to have been the cause of the large flight that appeared here so recently as the last days in April," 1878. This flock was mentioned in the 'Star' of April the 27th as follows:—"A countryman informs us that a few days since, whilst he was at L'ancresse Common, he saw several flocks of these smallest of British birds, numbering many hundreds in each, settle in different parts of the Common before dispersing over the Island. In verification of his words he showed us two or three of these tiny songsters which he had succeeded in knocking down with a stick." This large migratory flock had entirely disappeared from L'ancresse Common when we went to live there for two months in May of the same year; there was not then a Golden Crest to be seen about the Common. The whole flock had probably resumed their journey together, none of them having "dispersed over" or remained in the Island, and certainly, as far as I could judge, the numbers in other parts of the Island had not increased beyond what was usual and one might ordinarily expect. I have not been able to learn that the migratory flock above spoken of extended to any of the other Islands.

The Golden-crested Wren is mentioned by Professor Ansted, and marked as occurring in Guernsey and Sark. There are two—a male and female—in the Museum.

44. FIRE-CRESTED WREN. *Regulus ignicapillus*, C.L. Brehm. French, "Roitelet a triple bandeau."—I have a pair of these killed in Guernsey about 1872, but I have not the exact date; and Mr. Couch, who knew the Fire-crested Wren well, writing to me on the 23rd of March, 1877, says:—"I had the head and part of a Fire-crest female brought me by a young lady. She told me her brother knocked down two, and the other had a beautiful red and gold crest; so it must have been the male." As Mr. Couch knew both the Goldcrest and Fire-crest well, and the distinction between them, I have no doubt he rightly identified the bird which was brought to him. These and the pair in my collection are the only Guernsey specimens I can be certain of.

The 'Star' newspaper, however, in the note above quoted as to the migratory flock of Golden-crests, says:—"It may be a fact hitherto unknown to many of our readers that the Fire-crested Wren, very similar in appearance to the Golden-crested Wren, is not very uncommon in our Island. The Fire-crested Wren so closely resembles its *confrhre*, the Golden-crested Wren, that only a practised eye can distinguish the difference between them." I do not quite agree with the 'Star' as to the Fire-crest not being "very uncommon," though it occasionally occurs. I do not think it can be considered as anything but a rare occasional straggler. And this from its geographical distribution, which is rather limited, is what one would expect; it is not very common on the nearest coast of France or England, though it occasionally occurs about Torbay, which is not very far distant.

The name Fire-crest has probably led to many mistakes between this bird and the Golden-crest, as a brightly-coloured male Gold-crest has the golden part of the crest quite as bright and as deeply coloured as the Fire-crest; and the female Fire-crest has a crest not a bit more deeply coloured than the female Gold-crest. In point of fact the colour of the crest is of no value whatever in distinguishing between the birds, and the "practised eye" would find itself puzzled if it only relied upon that.

The French name for the Fire-crest, however, "Roitelet ` triple bandeau," is much more descriptive, as under the golden part of the crest there is a streak of black, and under that again a streak of white over the eye, and a streak of black through the eye; there is also a streak, or rather perhaps a spot of white, under the eye. The Gold-crest has only the streak of black immediately under the gold crest; below that the whole of the side of the face and the space immediately surrounding the eye is a uniform dull olive-green. If this distinction is once known and attended to the difference between the two birds may be immediately detected by even the unpractised eye.

A very interesting account of the nesting of this bird is given by Mr. Dresser, in his 'Birds of Europe,' he having made a journey to Altenkirchen, where the Fire-crest is numerous, on purpose to watch it in the breeding-season. The nest he describes as very like that of the Golden-crest; the eggs also are much like those of that bird, though a little redder in colour.

The Fire-crest is not mentioned in Professor Ansted's list, and there is no specimen at present in the Museum.

45. WREN. *Troglodytes parvulus*, K.L. Koch. French, "Roitelet," "Troglodyte mignon," "Troglodyte ordinaire."—The Wren is common and resident in all the Islands, and very generally distributed, being almost as common amongst the wild rocks on the coast as in the inland parts. On the 7th of July, 1878, I found a Wren's nest amongst some of the wildest rocks in the Island; the hinder part of the nest was wedged into a small crevice in the rock very firmly, the nest projecting and apparently only just stuck against the face of the rock. A great deal of material had been used, and the nest, projecting from the face of the rock as it did, looked large, and when I first caught sight of it I thought I might have hit upon an old Water Ouzel's nest. On getting close, however, I found it was only a Wren's, with young birds in it. I visited this nest several times, and saw the old bird feeding her young. I could not, however, quite make out what she fed them with, but I think with insects caught amongst the seaweed and tangle amongst the rocks. After the young were flown I took this nest, and was astonished to find, when it was taken out of the crevice, how much material had been used in wedging it in, and how firmly it was attached to the rock. This was certainly necessary to keep it in its place in some of the heavy gales that sometimes happen even at that time of year; in a very heavy northwesterly gale it would hardly have been clear of the wash of the waves at high water.

The Wren is included in Professor Ansted's list, but marked as only occurring in Guernsey. There is no specimen in the Museum.

46. TREE-CREEPER. *Certhia familiaris*, Linnaeus. French, "Grimpereau," "Grimpereau familier."—The Tree-creeper is resident and not uncommon in all the Islands, except perhaps Alderney, in

which Island I have never seen it. In Guernsey it may be seen in most of the wooded parts, and frequently near the town, in the trees on the lawns at Candie, Castle Carey, and in the New Ground. I have never seen it take to the rocks near the sea, like the Wren.

It is mentioned in Professor Ansted's list, and marked as occurring in Guernsey and Sark. There is no specimen in the Museum.

47. GREAT TIT. *Parus major*, Linnaeus. French, "Misange Charbonnihre."—The Paridae are by no means well represented in the Islands, either individually or as to number of species; and the Guernsey gardeners can have very little cause to grumble at damage done to the buds by the Tits. The Great Tit is moderately common and resident in Guernsey, but by no means so common as in England. During the whole two months I was in the Island this last summer, 1878, I only saw two or three Great Tits, and this quite agrees with my experience in June and July, 1866, and at other times.

The Great Tit is included in Professor Ansted's list, but only marked by him as occurring in Sark.

48. BLUE TIT. *Parus caeruleus*, Linnaeus. French, "Misange bleue."—Like the Great Tit, the Blue Tit is resident in all the Islands, but by no means numerous. In Guernsey it is pretty generally distributed over the more cultivated parts, but nowhere so numerous as in England. It is included in Professor Ansted's list, and marked as occurring in Guernsey and Sark.

I have not included either the Cole Tit or the Marsh Tit in this list, as I have never seen either bird in the Islands, and have not been able to find that they are at all known either in Guernsey or any of the other Islands.

Professor Ansted, however, includes the Cole Tit in his list, and marks it as occurring in Guernsey, but no other information whatever is given about it; and there is no specimen in the Museum, as there is of both the Great and the Blue Tits. I have not succeeded in getting a specimen myself.

49. LONG-TAILED TIT. *Acredula caudata*, Linnaeus. French, "Masange ` longue queue." [10] — The Long-tailed Tit is certainly far from common in Guernsey at present, and I have never seen it in the Islands myself. But Mr. MacCulloch writes me word—"The Long-tailed Tit is, or at least was, far from uncommon. Probably the destruction of orchards may have rendered it less common. The nest was generally placed in the forked branch of an apple-tree, and so covered with grey lichens as to be almost indistinguishable. I remember, in my youth, finding a nest in a juniper-bush."

It is included in Professor Ansted's list, and marked as occurring in Guernsey and Sark. There is, however, no specimen now in the Museum.

I am very doubtful as to whether I ought to include the Bearded Tit, *Panurus biarmicus* of Linnaeus, in this list. There are a pair in the Museum, but these may have been obtained in France or England. One of Mr. De Putron's men, however, described a bird he had shot in the reeds in Mr. De Putron's pond in the Vale, and certainly his description sounded very much as if it had been a Bearded Tit; but the bird had been thrown away directly after it was shot, and there was no chance of verifying the description.

50. WAXWING. *Ampelis garrulus*, Linnaeus. French, "Jaseur de Bohjme," "Grand Jaseur." — As would seem probable from its occasional appearance in nearly every county in England, the Waxwing does occasionally make its appearance in Guernsey as a straggler. I have never seen it myself, but Mr. MacCulloch writes me word—"I have known the Bohemian Waxwing killed here on several occasions, but have not the date."

An interesting account of the nesting habits of this bird, and the discovery of the nests and eggs by Mr. Wolley, was published by Professor Newton in the 'Ibis' for 1861, and will be found also in

Dresser's 'Birds of Europe.' and in the new edition of 'Yarrell,' by Professor Newton.

It is included in Professor Ansted's list, and marked as occurring in Guernsey; and there is one specimen in the Museum.

51. PIED WAGTAIL. *Motacilla lugubris*, Temminck. French, "Bergeronette Yarrellii." [11] — The Pied Wagtail has probably been better known to some of my readers as *Motacilla Yarrellii*, but, according to the rules of nomenclature before alluded to, *Motacilla lugubris* of Temminck seems to have superseded the probably better-known name of *Motacilla Yarrellii*.

For some reason or other the Pied Wagtail has grown much more scarce in Guernsey than it used to be; at one time it was common even about the town, running about by the gutters in the street, and several were generally to be seen on the lawn at Candie. But this last summer—that of 1878—I did not see one about Candie, or indeed anywhere else, except one pair which were breeding near the Vale Church; and when there in November, 1875, I only saw one, and that was near Vazon Bay. Mr. MacCulloch has also noticed this growing scarcity of the Pied Wagtail, as he writes to me—"Of late years, for some reason or other, Wagtails of all sorts have become rare." In the summer of 1866, however, I found the Pied Wagtail tolerably common.

It is included in Professor Ansted's list, and marked as occurring in Guernsey and Sark.

52. WHITE WAGTAIL. *Motacilla alba*, Linnaeus. French, "Lavendihre," "Hoche-queue grise," "Bergeronette grise."—The White Wagtail is still scarcer than the Pied, but I saw one pair evidently breeding between L'ancresse Road and Grand Havre. The White Wagtail so much resembles the Pied Wagtail, that it may have been easily overlooked, and may be more common than is generally known.

The fully adult birds may easily be distinguished, especially when in full breeding plumage, as the back of the Pied Wagtail is black, while that of the White Wagtail is grey. After the autumnal moult, however, the distinction is not quite so easy, as the feathers of the Pied Wagtail are then margined with grey, which rather conceals the colour beneath; but if the feathers are lifted up they will be found to be black under the grey margins. The young birds of the year, in their first feathers, cannot be distinguished, and the same may be said of the eggs.

The White Wagtail is included in Professor Ansted's list, but marked as only occurring in Guernsey. There is no specimen either of the Pied or White Wagtail in the Museum.

53. GREY WAGTAIL. *Motacilla melanope*, Pallas. French, "Bergeronette jaune." — The Grey Wagtail is by no means common in the Islands, though it may occasionally remain to breed, as I have seen it both in Guernsey and Sark between the 21st of June and the end of July in 1866, but I have not seen it in any of the Islands during the autumn. It is, however, no doubt an occasional, though never very numerous, winter visitant, probably more common, however, at this time of year than in the summer, as I have one in winter plumage shot in Guernsey in December, and another in January, 1879, and there is also one in the Museum in winter plumage.

Professor Ansted includes it in his list, and marks it as occurring in Guernsey and Sark.

54. YELLOW WAGTAIL. *Motacilla raii*, Bonaparte. French, "Bergeronnette flaviole." — As far as I have been able to judge the Yellow Wagtail is only an occasional visitant on migration. A few, however, may sometimes remain to breed. I have one Channel Island specimen killed in Guernsey the last week in March. Mr. MacCulloch, however, writes me word that in some years they — *i.e.*, Yellow Wagtails — are not very uncommon, but of late, for some reason or other, Wagtails of all sorts have become rare. He adds — "I am under

the impression that we have more than one Yellow Wagtail." It is, therefore, possible that the Greyheaded Wagtail, the true *Motacilla flava* of Linnaeus, may occasionally occur, or in consequence of the bright yellow of portions of its plumage the last-mentioned species — the Grey Wagtail — may have been mistaken for a second species of Yellow Wagtail. I have not myself seen the Yellow Wagtail in either of the Islands during my summer visits in 1866, 1876, or 1878; so it certainly cannot be very common during the breeding-season, or I could scarcely have missed seeing it.

Professor Ansted has not included it in his list, and there is no specimen at present in the Museum.

55. TREE PIPIT. *Anthus trivialis*, Linnaeus. French, "Pipit des arbres," "Pipit des buissons." — A very numerous summer visitant to all the Islands, breeding in great numbers in the parts suited to it. In the Vale it was very common, many of the furze-bushes on L'Ancresse Common containing nests. The old male might constantly be seen flying up from the highest twigs of the furze-bush, singing its short song as it hovered over the bush, and returning again to the top branch of that or some neighbouring bush. This continued till about the middle of July, when the young were mostly hatched, and many of them flown and following their parents about clamorous for food, which was plentiful in the Vale in the shape of numerous small beetles, caterpillars, and very small snails. The young were mostly hatched by the beginning of July, but I found one nest with young still in it in a furze-bush about ten yards from high watermark as late as the 27th of July, but the young were all flown when I visited the nest two days afterwards. The Tree Pipits have all departed by the middle of October, and I have never seen any there in November.

The Tree Pipit is mentioned in Professor Ansted's list, but no letters marking the distribution of the species amongst the Islands are given. There is no specimen of this or either of the other Pipits in the Museum.

56. MEADOW PIPIT. *Anthus pratensis*, Linnaeus. French, "Le cujelier," "Pipit des pris," "Pipit Farlouse."—The Meadow Pipit is resident and breeds in all the Islands, but is by no means so numerous as the Tree Pipit is during the summer. I think, however, its numbers are slightly increased in the autumn, about the time of the departure of the Tree Pipits, by migrants.

It is included by Professor Ansted in his list, but marked as occurring only in Guernsey.

57. ROCK PIPIT. *Anthus obscurus*, Latham. French, "Pipit obsur," "Pipit spioncelle."—Resident and numerous, breeding amongst the rocks and round the coast of all the Islands. It is also common in all the small outlying Islands, such as Burhou, and all the little rocky Islands that stretch out to the northward of Herm, and are especially the home of the Puffin and the Lesser Black-backed Gull. On all of these the Rock Pipit may be found breeding, but its nest is generally so well concealed amongst the thrift samphire, wild stock, and other seaside plants which grow rather rankly amongst those rocks, considering how little soil there generally is for them and what wild storms they are subject to, that it is by no means easy to find it, though one may almost see the bird leave the nest.

The Bock Pipit is included in Professor Ansted's list, but marked as only occurring in Guernsey. All the Rock Pipits I have seen in the Channel Islands have been the common form, *Anthus obscurus*; I have never seen one of the rufous-breasted examples which occur in Scandinavia and the Baltic, and have by some been separated as a distinct species under the name of *Anthus rupestris*.

58. SKY LARK. *Alauda arvensis*, Linnaeus. French, "Alouette des champs."—Mr. Mitivier, in his 'Dictionary,' gives Houhdre as the local Guernsey-French name of the Sky Lark. As may be supposed by its having a local name, it is a common and well-known bird,

and is resident in all the Islands. I have not been able to find that its numbers are much increased by migrants at any time of year, though probably in severe weather in the winter the Sky Larks flock a good deal, as they do in England. The Sky Lark breeds in all the Islands, and occasionally places its nest in such exposed situations that it is wonderful how the young escape. One nest we found by a roadside near Ronceval; it was within arm's length of the road, and seemed exposed to every possible danger. When we found it, on the 15th of June, there were five eggs in it, fresh, or, at all events, only just sat on, as I took one and blew it for one of my daughters. On the 19th we again visited the nest; there were then four young ones in it, but they were so wonderfully like the dry grass which surrounded the nest in colour that it was more difficult to find it then than when the eggs were in it, and except for the young birds moving as they breathed I think we should not have found it a second time. A few days after — July the 3rd — there was very heavy rain all night. Next day we thought the Sky Larks must be drowned (had they been Partridges under the care of a keeper they would have been), but as it was only one was washed out of the nest and drowned; the rest were all well and left the nest a few days after. So in spite of the exposed situation close to a frequented road, on a bit of common ground where goats and cows were tethered, nets and seaweed, or "vraic," as it is called in Guernsey, spread for drying, dogs, cats, and children continually wandering about, and without any shelter from rain, the old birds brought off three young from their five eggs.

The Sky Lark is mentioned in Professor Ansted's list as occurring only in Guernsey and Sark. It is, however, quite as common in Alderney and Herm. There is no specimen in the Museum.

59. SNOW BUNTING. *Plectrophanes nivalis*, Linnaeus. French, "Ortolan de neige," "Bruant de neige." — The Snow Bunting is probably a regular, though never very numerous, autumnal visitant, remaining on into the winter. It seems to be more numerous in some years than others. Mr. Mac Culloch tells me a good many Snow Buntings were seen in November, 1850.

Mr. Couch records one in the 'Zoologist' for 1874 as having been killed at Cobo on the 28th of September of that year. This seems rather an early date. When I was in Guernsey in November, 1875, I saw a few flocks of Snow Buntings, and one—a young bird of the year—which had been killed by a boy with a catapult, was brought into Couch's shop about the same time, and I have one killed at St. Martin's, Guernsey, in November, 1878; and Captain Hubbach writes me word that he shot three out of a flock of five in Alderney in January, 1863.

Professor Ansted mentions the Snow Bunting in his list as occurring in Guernsey and Sark, and there is a specimen at present in the Museum.

60. BUNTING. *Emberiza miliaria*, Linnaeus. French, "Le proyer," "Bruant proyer."—The Bunting is resident in Guernsey and breeds there, but in very small numbers, and it is very local in its distribution. I have seen a few in the Vale. I saw two or three about the grounds of the Vallon in July, 1878, which were probably the parents and their brood which had been hatched somewhere in the grounds.

It is mentioned in Professor Ansted's list as occurring only in Guernsey. There is one specimen in the Museum.

61. YELLOW HAMMER. *Emberiza citrinella*, Linnaeus. French, "Bruant jaune."—The Yellow Hammer, though resident and breeding in all the Islands, is by no means as common as in many parts of England. In Alderney perhaps it is rather more common than in Guernsey, as I saw some near the Artillery Barracks this summer, 1878, and Captain Hubbach told me he had seen two or three pairs about there all the year. In Guernsey, on the other hand, I did not see one this summer, 1878. I have, however, shot a young bird there which certainly could not have been long out of the nest. I have never seen the Cirl Bunting in any of the Islands, nor has it, as far as I know, been recorded from them, which seems rather surprising, as it is common on the South Coast of Devon, and migratory, but not numerous, on the North Coast of France; [12] so it is very probable that it may yet occur.

The Yellow Hammer is included in Professor Ansted's list, and marked as occurring in Guernsey and Sark. There are also a pair in the Museum.

62. CHAFFINCH. *Fringilla caelebs*, Linnaeus. French, "Pinson ordinaire," "Grosbec pinson."—- The Chaffinch is resident, tolerably common, and generally distributed throughout the Islands, but is nowhere so common as in England. In Guernsey this year, 1878, it seemed to me rather to have decreased in numbers, as I saw very few,—certainly not so many as in former years,—though I could not find that there was any reason for the decrease.

It is, of course, mentioned in Professor Ansted's list, but by him only marked as occurring in Guernsey and Sark. There is only one—a female—at present in the Museum.

63. BRAMBLING. *Fringilla montifringilla*, Linnaeus. French, "Pinson d'Ardennes." "Grosbec d'Ardennes."—The Brambling can only be considered an occasional autumn and winter visitant, and probably never very numerous. I have never seen the bird in the Channel Islands myself. I have, however, one specimen—a female—killed in Brock Road, Guernsey, in December, 1878, and I have been informed by Mr. MacCulloch that he had a note of the occurrence of the Brambling or Mountain Finch in January, 1855. It cannot, however, be looked upon as anything more than a very rare occasional straggler, by no means occurring every year.

It is mentioned in Professor Ansted's list, and marked as occurring in Guernsey and Sark. There is no specimen at present in the Museum.

64. TREE SPARROW. *Passer montanus*, Linnaeus. French, "Friquet."—The Tree Sparrow breeds, and is probably resident in the Islands. Up to this year, 1878, I have only seen it once myself, and

that was on the 7th of June, 1876, just outside the grounds of the Vallon in Guernsey. From the date and from the behaviour of the bird I have no doubt it had a nest just inside the grounds. I could not then, however, make any great search for the nest without trespassing, though I got sufficiently near the bird to be certain of its identity. This year, 1878, I could not see one anywhere about the Vallon, either inside or outside the grounds. I saw, however, one or two about the Vale, but they were very scarce. I have not myself seen the Tree Sparrow in any of the other Islands.

It is included in Professor Ansted's list, and marked as occurring in Sark only. I have not seen a specimen at Mr. Couch's, or any of the other bird-stuffers, but there is one in the Museum and some eggs, all of which are probably Guernsey.

65. HOUSE SPARROW. *Passer domesticus*, Linnaeus. French, "Moineau domestique," "Grosbec moineau."—The House Sparrow is very numerous throughout the Islands, abounding where there are any buildings inhabited by either man, horses, or cattle. In the gardens near the town of St. Peter's Port, in Guernsey, it is very common, and does a considerable amount of mischief. It is, however, by no means confined to the parts near the town, as many were nesting in some ilex trees near the house we had on L'Ancresse Common, although the house had been empty since the previous summer, and the garden uncultivated; so food till we came must have been rather scarce about there. As the wheat is coming into ear the Sparrows, as in England, leave the neighbourhood of the town and other buildings and spread themselves generally over the country, for the purpose of devouring the young wheat while just coming into ear and still soft. In Alderney, owing probably in a great measure to the absence of cottages, farm-buildings, and stables at a distance from the town, and also perhaps owing to the absence of hedges, it is not so numerous in the open part, and consequently not so mischievous, being mostly confined to the town, and to the buildings about the harbour-works. The young wheat, however, is still a temptation, and is accordingly punished by the Sparrows.

The House Sparrow is mentioned by Professor Ansted in his list, but no letters are given marking the general distribution over the Islands, probably because it is so generally spread over them. The local Guernsey-French name is "Grosbec," for which see Mitivier's 'Dictionary.'

66. HAWFINCH. *Coccothraustes vulgaris*, Pallas. French, "Grosbec." — The Hawfinch or Grosbeak, as it is occasionally called, is by no means common in Guernsey, and I have never seen it there myself, but I have a skin of one killed in the Catel Parish in December, 1878; and Mr. MacCulloch informs me it occasionally visits that Island in autumn, but in consequence of its shy and retiring habits it has probably been occasionally overlooked, and escaped the notice of the numerous gunners to whom it would otherwise have more frequently fallen a victim. The bird-stuffer and carpenter in Alderney had one spread out on a board and hung up behind his door, which had been shot by his friend who shot the Greenland Falcon, in the winter of 1876 and 1877, somewhere about Christmas. I know no instance of its remaining to breed in the Islands, though it may occasionally do so in Guernsey, as there are many places suited to it, and in which it might well make its nest without being observed. As it seems increasing in numbers throughout England, it is by no means improbable that it will visit the Channel Islands more frequently. The Hawfinch is included in Professor Ansted's list, and by him marked as occurring only in Guernsey. There are two specimens in the Museum.

67. GREENFINCH. *Coccothraustes chloris*, Linnaeus. French, "Grosbec verdier," "Verdier ordinaire." — The Greenfinch is a common resident, and breeds in all the Islands, but is certainly not quite so common as in England. It is more numerous perhaps in Guernsey and Sark than in Alderney; it is also pretty common in Jethou and Herm.

It is included in Professor Ansted's list, but only marked as occurring in Guernsey and Sark. There is no specimen in the Museum.

68. GOLDFINCH. *Carduelis elegans*, Stephens. French, "Chardonneret," "Grosbec chardonneret." — The Goldfinch is resident in and breeds in all the Islands. In Guernsey I was told a few years ago that it had been much more numerous than it then was, the bird-catchers having had a good deal to answer for in having shortened its numbers. It is now, however, again increasing its numbers, as I saw many more this year (1878) than I had seen before at any time of year. There were several about the Grand Mare, and probably had nests there, and I saw an old pair, with their brood out, at St. George on the 5th of June, and soon after another brood about Mr. De Putron's pond, where they were feeding on the seeds of some thistles which were growing on the rough ground about the pond. I have also seen a few in Alderney; and Captain Hubbach writes me word that the Goldfinch was quite plentiful here (Alderney) in the winter of 1862 and 1863. But he adds — "I have not seen one here this year." So probably its numbers are occasionally increased by migratory flocks in the winter.

Professor Ansted includes the Goldfinch in his list, but marks it as occurring only in Guernsey and Sark. There is no specimen in the Museum.

69. SISKIN. *Carduelis spinus*, Linnaeus. French, "Tarin," "Grosbec tarin." — The Siskin can only be looked upon as an occasional, accidental visitant — indeed, I only know of one instance of its occurrence, and that is recorded by Mr. Couch at p. 4296 of the 'Zoologist' for 1875 in the following words: — "I have the first recognised specimen of the Siskin; a boy knocked it down with a stone in an orchard at the Vrangue in September." This communication is dated November, 1874. I have never seen the Siskin in any of the Channel Islands myself, and Mr. MacCulloch writes me word — "I have never heard of a Siskin here, but, being migratory, it may occur." I see,

however, no reason to doubt Mr. Couch's statement in the 'Zoologist,' as the bird was brought into his shop. He must have had plenty of opportunity of identifying it, though he does not tell us whether he preserved it. There can, however, be no possible reason why the Siskin should not occasionally visit Guernsey on migration, as it extends its southern journey through Spain to the Mediterranean and across to the North-western Coast of Africa; and the Channel Islands would seem to lie directly in its way.

The Siskin, however, is not mentioned in Professor Ansted's list, and there is no specimen at present in the Museum.

70. LINNET. *Linota cannabina*, Linnaeus. French, "Linotte," "Grosbec linotte."—The Linnet is resident and the most numerous bird in the Islands by far, outnumbering even the House Sparrow, and it is equally common and breeds in all the Islands. The Channel Islands Linnets always appear to me extremely bright-coloured, the scarlet on the head and breast during the breeding-season being brighter than in any British birds I have ever seen. Though the Linnet is itself so numerous, it is, as far as I have been able to ascertain, the only representative of its family to be found in the Channel Islands; at least I have never seen and had no information of the occurrence of either the Lesser Redpole, the Mealy Redpole, or the Twite, though I can see no reason why each of these birds should not occasionally occur.

The Linnet is included in Professor Ansted's list, but marked by him as only occurring in Guernsey and Sark; and there is a specimen in the Museum.

71. BULLFINCH. *Pyrrhula europaea*, Vieillot. French, "Bovreuil commun."—Miss C.B. Carey, in the 'Zoologist' for 1874, mentions a Bullfinch having been brought into Couch's shop in November of that year, and adds—"This bird is much more common in Jersey than it is here." Miss Carey is certainly right as to its not being common in Guernsey, as I have never seen the bird on any of my

expeditions to that Island, nor have I seen it in either of the other Islands which come within my district.

Professor Ansted includes the Bullfinch in his list, but oddly enough only marks it as occurring in Guernsey and Sark, although Mr. Gallienne, in his remarks published with the list, says—"The Bullfinch occasionally breeds in Jersey, but is rarely seen in Guernsey," so far agreeing with Miss Carey's note in the 'Zoologist,' but he does not add anything about Sark. There is no specimen in the Museum.

72. COMMON CROSSBILL. *Loxia curvirostra*, Linnaeus. French, "Bec-croisi," "Bec-croisi commun."—The Crossbill is an occasional visitant to all the Islands, and sometimes in considerable numbers, but, as in England, it is perfectly irregular as to the time of year it chooses for its visits. Mr. MacCulloch writes me word—"The Crossbill is most uncertain in its visits. Many years will sometimes pass without a single one being heard of. When they do come it is generally in large flocks. I have known them arrive in early autumn, and do great havoc amongst the apples, which they cut up to get at the pips. Sometimes they make their appearance in the winter, seemingly driven from the Continent by the cold."

My first acquaintance with the Crossbill was in Sark on the 25th of June, 1866, when I saw a very fine red-plumaged bird in a small fir-plantation in the grounds of the Lord of Sark. It was very tame, and allowed me to approach it very closely. I did not see any others at that time amongst the fir-trees, though no doubt a few others were there. On my return to Guernsey on the following day I was requested by a bird-catcher to name some birds that were doing considerable damage in the gardens about the town. Thinking from having seen the one in Sark, and from his description, that the birds might be Crossbills, I asked him to get me one or two, which he said he could easily do, as the people were destroying them on account of the damage they did. In a day or two he brought me one live and two dead Crossbills, and told me that as many as forty had been shot in one person's garden. The two dead ones he brought me were one in red and the other in green plumage, and the live one was in green plumage. This one I brought home and kept in my aviary till March, 1868, when it was killed by a Hawk striking it through the wires. It was, however, still in the same green plumage when it was killed as it was when I brought it home, though it had moulted twice.

The Crossbill did not appear at that time to be very well known in Guernsey, as neither the bird-catcher nor the people in whose gardens the birds were had ever seen them before or knew what they were. This year (1866), however, appears to have been rather an exceptional year with regard to Crossbills, as I find some recorded in the 'Zoologist' from Norfolk, the Isle of Wight, Sussex, and Henley-on-Thames, about the same time; therefore there must have been a rather widely-spread flight. From that time I did not hear any more of Crossbills in the Islands till December, 1876, when Mr. Couch sent me a skin of one in reddish plumage, writing at the same time to say—"The Crossbill I sent from its being so late in the season when it was shot—the 11th of December; there were four of them in a tree by Haviland Hall. I happened to go into the person's house who shot it, and his children had it playing with."

I do not know that there is any evidence of the Crossbill ever having bred in the Islands, though it seems to have made its appearance there at almost all times of year. Mr. MacCulloch mentions its feeding on the apple-pips, and doing damage in the orchards accordingly, and I know it is generally supposed to do so, and has in some places got the name of "Shell Apple" in consequence, but though I have several times kept Crossbills tame, and frequently tried to indulge them with apples and pips, I have never found them care much about them; and a note of Professor Newton's, in his edition of 'Yarrell,' seems to agree with this. He says:—"Of late it has not been often observed feeding on apples, very possibly owing to the greatly-increased growth of firs, and especially larches, throughout the country. In Germany it does not seem ever to have been known as attacking fruit-trees."

The Crossbill is included in Professor Ansted's list, and only marked as occurring in Guernsey and Sark. There is no specimen in the Museum.

73. COMMON STARLING. *Sturnus vulgaris*, Linnaeus. French, "Etourneau vulgaire." — The Starling is sometimes very numerous in the autumn, but those remaining throughout the year and breeding in the Island are certainly very few in number, as I have never seen the Starling in any of my summer visits; and Mr. MacCulloch tells me "the Starling may possibly still breed here, but it certainly is not common in summer. A century ago it used to nest in the garrets in the heart of the town." As to its not being common in summer, that quite agrees with my own experience, but a few certainly do breed in the Island still, or did so within a very few years, as Miss C.B. Carey had eggs in her collection taken in the Island in 1873 or 1874, and I have seen eggs in other Guernsey collections, besides those in the Museum. When I was in Guernsey in November, 1871, Starlings were certainly unusually plentiful, even for the autumn, very large flocks making their appearance in all parts of the Island, and in the evening very large flocks might be seen flying and wheeling about in all directions before going to roost. Many of these flocks I saw fly off in the direction of Jersey and the French coast, and they certainly continued their flight in that direction as long as I could follow them with my glass, but whether they were only going to seek a roosting-place and to return in the morning, or whether they continued their migration and their place was supplied by other flocks during the night, I could not tell, but certainly there never seemed to be any diminution in their numbers during the whole time I was there from the 1st to the 16th of November. I think it not at all improbable that many of these flocks only roosted out of the Island and returned, as even here in Somerset they collect in large flocks before going to roost, and fly long distances, sometimes quite over the Quantock Hills, to some favourite roosting-place they have selected, and return in the morning, and the distance would in many places be nearly as great. These flocks of Starlings seem to have continued in the Island quite into the winter, as Miss Carey notes, in the 'Zoologist' for 1872, seeing a flock in the field before the house at Candie close to the town as late as the 6th of December, 1871. At the same time that there were so many in Guernsey, Starlings were

reported as unusually numerous in Alderney, but how long the migratory flocks remained there I have not been able to ascertain.

The Starling is included in Professor Ansted's list, but marked as only occurring in Guernsey and Sark. There are two specimens in the Museum and some eggs.

74. CHOUGH. *Pyrrhocorax graculus*, Linnaeus. French, "Crave." — The Chough is a common resident in Guernsey, breeding amongst the high rocks on the south and east part of the Island, and in the autumn and winter spreading over the cultivated parts of the Island, sometimes in considerable flocks, like Rooks.

As Jackdaws are by no means numerous in Guernsey, and as far as I have been able to make out never breed there, the Choughs have it all their own way, and quite keep up their numbers, even if they do not increase them, which I think very doubtful, though I can see no reason why they should not, as their eggs are always laid in holes in the cliffs, and very difficult to get at, and at other times of year the birds are very wary, and take good care of themselves, it being by no means easy to get a shot at them, unless by stalking them up behind a hedge or rock; and as they are not good eating, and will not sell in the market like Fieldfares and Redwings, no Guernsey man thinks of expending powder and shot on them; so though not included in the Guernsey Bird Act, the Choughs on the whole have an easy time of it in Guernsey, and ought to increase in numbers more than they apparently do. In Sark the Choughs have by no means so easy a time, as the Jackdaws outnumber them about the cliffs, and will probably eventually drive them out of the Island — indeed, I am afraid they have done this in Alderney, as I did not see any when there in the summer of 1876, nor in this last summer (1878); and Captain Hubbach writes me word he has seen none in Alderney himself this year (1878). I, however, saw some there in previous visits, but now for some reason, probably the increase of Jackdaws, the Choughs appear to me nearly, if not quite, to have deserted that Island. In Herm and Jethou there are also a few Choughs, but Jackdaws are the more numerous in both Islands. No Choughs appear to inhabit the small rocky islets to the northward of Herm, though some of them appear to be large enough to afford a breeding-place for either Choughs or Jackdaws, but neither of these birds seem to have taken possession of them; probably want of food is the occasion of this. Mr. Mitivier, in his 'Rimes Guernseaise,' gives "Cahouette" as the local Guernsey-French name of the Chough, though I suspect the name is equally applicable to the Jackdaw.

The Chough is mentioned in Professor Ansted's list, but marked as only occurring in Guernsey and Sark. There are two specimens in the Museum.

75. JACKDAW. *Corvus monedula*, Linnaeus. French, "Choucas," "Choucas gris."—I am quite aware that many Guernsey people will tell you that there are no Jackdaws in Guernsey, but that their place is entirely taken by Choughs. Mr. MacCulloch seems to be nearly of this opinion, as he writes me—"I suppose you are right in saying there are a few Jackdaws in Guernsey, but I do not remember ever to have seen one here;" and he adds—"I believe they are common in Alderney," which is certainly the case; as I said above, they have almost, if not quite, supplanted the Choughs there. There are, however, certainly a few Jackdaws in Guernsey, as I have seen them there on several occasions, but I cannot say that any breed there, and I think they are only occasional wanderers from the other Islands, Sark, Jethou, and Herm, where they do breed. Mr. Gallienne's note to Professor Ansted's list seems to agree very much with this, as he says—"The Jackdaw, which is a regular visitor to Alderney, is rarely seen in Guernsey." It is now, however, resident in Alderney, as well as in Sark, Jethou, and Herm.

It is mentioned in Professor Ansted's list, but only marked as occurring in Guernsey and Sark, nothing being said about Alderney and the other Islands in spite of Mr. Gallienne's note. There is no specimen at present in the Museum.

76. RAVEN. *Corvus corax*, Linnaeus. French, "Corbeau," "Corbeau noir."—The Raven can now only be looked upon as an occasional straggler. I do not think it breeds at present in any of the Islands, as I have not seen it anywhere about in the breeding-season since 1866, when I saw a pair near the cliffs on the south-end of the Island in June; but as the Raven is a very early breeder, these may have only been wanderers. It is probably getting scarcer in Guernsey, as I have not seen any there since; and the last note I have of Ravens being seen in the Island is in a letter from Mr. Couch, who wrote me word that two Ravens had been seen and shot at several times, but not obtained, in November, 1873. I have not seen a Raven in any of the other Islands, and do not know of one having occurred there.

Professor Ansted includes it in his list, and marks it as only occurring in Guernsey. There is no specimen at present in the Museum.

77. CROW. *Corvus corone*, Linnaeus. French, "Corneille noire."— The Crow is pretty common, and breeds in most of the Islands, and probably at times commits considerable depredations amongst the eggs and young of the Gulls and Shags—at all events it is by no means a welcome visitor to the breeding stations of the Gulls, as in this summer (1878) I saw four Crows about a small gullery near Petit Bo Bay, one of which flew over the side of the cliff to have a look at the Gulls' eggs, probably with ulterior intentions in regard to the eggs; but one of the Gulls saw him, and immediately flew at him and knocked him over: what the end of the fight was I could not tell, but probably the Crow got the worst of it, as several other Gulls went off to join their companion as soon as they heard the row; and the Crows trespassed no more on the domain of the Gulls—at least whilst I was there, which was some time.

Professor Ansted includes the Crow in his list, but only marks it as occurring in Guernsey and Sark. There is no specimen in the Museum.

78. HOODED CROW. *Corvus cornix*, Linnaeus. French, "Corbeau mantele," "Corneille mantelie."—The Hooded Crow can only be considered an occasional autumnal and winter visitant. I have never seen it myself in the Islands, though many of my visits to Guernsey have been in the autumn. Mr. Couch, however, reports a small flock of Hooded Crows being in Guernsey in November, 1873, one of which was obtained. Mr. MacCulloch writes me word that the Hooded Crow is a very rare visitant, and only, as far as he knows, in very cold weather; and he adds—"It is strange that we should see it so rarely, as it is very common about St. Maloes." Colonel l'Estrange, however, informed me that one remained in Sark all last summer—that of 1877—and paired with a common Crow, [13] but we could see nothing of the couple this year. I believe it is not at all uncommon for these birds to pair in Scotland and other places where both species are numerous in the breeding-season, but this is the only instance I have heard of in the Channel Islands—in fact, it is the only time I have heard of the Hooded Crow remaining on till the summer.

The Hooded Crow is included in Professor Ansted's list, and marked as occurring in Guernsey and Sark; and there are two specimens in the Museum.

79. ROOK. *Corvus frugilegus*, Linnaeus. French, "Freux", "Corbeau Freux."—I have never seen the Rook in the Islands myself, even as a stranger, but Mr. Gallienne in his notes to Professor Ansted's list, says, speaking of Guernsey, "The Rook has tried two or three times to colonise, but in vain, having been destroyed or frightened away." Mr. MacCulloch also writes me word much to the same effect, as he says "I have known Rooks occasionally attempt to build here (Guernsey), but they are invariably disturbed by boys and guns, and driven off. They sometimes arrive here in large flocks in severe winters."

The Rook is mentioned in Professor Ansted's list as occurring in Guernsey only, and there are two specimens in the Museum, both probably Guernsey killed.

80. MAGPIE. *Pica rustica*, Scopoli. French, "Pie", "Pie ordinaire." — The Magpie is resident and tolerably common in Guernsey, breeding in several parts of the Island; it is also resident, but I think not quite so common, in Sark. I do not remember having seen it in Alderney, and the almost entire absence of trees would probably prevent it being anything more than an occasional visitant to that Island.

It is included in Professor Ansted's list, but marked as only occurring in Guernsey; and there are two specimens in the Museum.

81. LESSER SPOTTED WOODPECKER. *Picus minor*, Linnaeus. French, "Pie ipeichette." — As may be expected, the Woodpeckers are not strongly represented in the Islands, and the present species, the Lesser Spotted Woodpecker, is the only one as to the occurrence of which I can get any satisfactory evidence.

Professor Ansted, however, includes the Greater Spotted Woodpecker in his list, and marks it as occurring in Guernsey only; and there is one specimen of the Green Woodpecker, *Gecinus viridis*, in the Museum, but there is no note whatever as to its locality; so under these circumstances I have not thought it right to include either species. But as to the occurrence of the Lesser Spotted Woodpecker, though I have not seen it myself, nor have I a Channel Island specimen, I have some more evidence; for in reply to some questions of mine on the subject, Mr. Couch wrote to me in April, 1877, "Respecting the Woodpecker, you may fully rely on the Lesser Spotted as having been shot here, four examples having passed through my hands; and writing from memory I will, as near as possible, tell you when and where they were shot. I took a shop here in 1866. In the month of August, 1867, there was one brought to me alive, shot in the water lanes, just under Smith's Nursery by a young gent at the College; he wounded it in the wing. I wanted too much to stuff it (2s. 6d.); he took the poor bird out, fixed it somewhere; he and his companions fired at it so often they blew it to atoms. The same year, early in September, one was shot at St. Martin's; I stuffed that for a lady: there were four in the same tree; the day following they were not to be found. The second week in October, the same year I had one, and stuffed it for the person who shot it out at St. Saviour's; there were two besides in the same tree, but I had neither one myself. In 1868, I stuffed one that was shot at St. Peter's, in December; it was taken home the Christmas Eve. These were all I have had, but I have heard of their being seen about since, twice or three times." In addition to this letter, which I have no reason to doubt, Mr. MacCulloch wrote me word—"We have in the Museum a Lesser Spotted Woodpecker, shot near Havilland Hall, in November, 1855; I saw it before it was stuffed." This bird was not in the Museum this year, (1878), as I looked everywhere for it, so I suppose it was moth-eaten and thrown away, like many others of the best specimens in the Museum, after the years of neglect they have been subject to. From these letters, there can be no doubt whatever that the Lesser Spotted Woodpecker has been occasionally procured in Guernsey, and that it may be considered either an occasional autumnal visitor, remaining on into winter, or, what is more probable, a thinly-scattered resident.

It is included in Professor Ansted's list, and marked as only occurring in Guernsey. As above stated, the specimen formerly in the Museum no longer exists.

82. WRYNECK. *Yunx torquilla*, Linnaeus. French, "Torcol ordinaire."—The Wryneck, or, as it is called in Guernsey-French, "Parlh" [14] is generally a numerous summer visitor to the Islands, arriving in considerable numbers, about the same time as the mackerel, wherefore it has also obtained the local name of "Mackerel Bird." It is generally distributed through the Islands, remaining through the summer to breed, and departing again in early autumn, August, or September. Its numbers, however, vary considerably in different years, as in some summers I have seen Wrynecks in almost every garden, hedgerow, or thick bush in the Island; always when perched, sitting across the branches or twigs, on which they were perched, and never longways or climbing, as would be the case with a Woodpecker or Creeper; and the noise made by the birds

during the breeding-season, was, in some years, incessant; this was particularly the case in the early part of the summer of 1866, when the birds were very numerous, and the noise made was so great that on one occasion I was told that the Mackerel Birds seriously interrupted a scientific game of *Croquet*, which was going on at Fort George, by the noise they made; I can quite believe it, as, though I was not playing in the game, I heard the birds very noisy in other parts of the Island. This last summer, however (1878), I saw very few Wrynecks—only four or five during the whole of the two months I was in the Islands, and hardly heard them at all.

It is included in Professor Ansted's list, but only marked as occurring in Guernsey and Sark. There are two specimens in the Museum.

83. HOOPOE. *Upupa Epops*, Linnaeus. French, "La Huppi," "Huppi ordinaire."—The Hoopoe, as may be supposed from its geographical range and from its frequent occurrence in various parts of England, is an occasional visitant to the Channel Islands during the seasons of migration, occurring both in spring and autumn with sufficient frequency to have gained the name of "Tuppe" in Guernsey-French. Though occurring in spring and autumn, I am not aware that it ever remains to breed, though perhaps it might do so if not shot on every possible occasion. This shooting of every straggler to the Channel Islands is a great pity, especially with the spring arrivals, as some of them might well be expected to remain to breed occasionally if left undisturbed; and the proof of the Hoopoe breeding in the Channel Islands would be much more interesting than the mere possession of a specimen of so common and well-known a bird: if a local specimen should be wanted, it could be obtained equally well in autumn, when there would be no question as to the breeding. The autumn arrivals seem also to be most numerous, at least judging from the specimens recorded during the last four or five years, as Mr. Couch records one, a female, shot near Ronseval, in Guernsey, on the 26th of September; and another also in Guernsey, shot on the 23rd of September; I have one, obtained in Alderney in August, though I have not the exact date; and another picked up

in a lane in St. Martin's parish, in Guernsey, on the 24th of August. During the same time I only know of one spring occurrence; that was on April the 10th of this year (1878), when two were seen, and one shot in Herm, as recorded in the 'Star' newspaper, for April the 13th; this one I saw soon afterwards at Mr. Jago's, the bird-stuffer. These birds were probably paired, and would therefore very likely have bred in Herm, had one of them not been shot, and the other accordingly driven to look for a mate elsewhere. It would pay, as well as be interesting, as I remarked in a note to the 'Star' in reference to this occurrence of the pair of Hoopoe's, to encourage these birds to breed in the Islands whenever they shewed a disposition to do so, as, though rather a foul-feeder and of unsavoury habits in its nest, and having no respect for sanitary arrangements, the Hoopoe is nevertheless one of the most useful birds in the garden.

The Hoopoe is included in Professor Ansted's list, but only marked as occurring in Guernsey and Sark. There are now only two specimens in the Museum, and these have no note of date or locality, but a few years ago there were several more, and one or two I remember were marked as having been killed in the spring; the rest were probably autumnal specimens.

84. CUCKOO. *Cuculus canorus*, Linnaeus. French, "Coucou gris."—The Cuckoo is one of the commonest and most numerous summer visitors to the Islands, and is generally spread over all of them; it arrives about the same time that it does in England, that is to say, about the middle of April. I know earlier instances—even as early as February—have been recorded, but these must have been recorded in consequence of some mistake, probably some particularly successful imitation of the note. Mr. MacCulloch seems to think that the time of their arrival is very regular, as he writes to me to say, "The Cuckoo generally arrives here about the 15th of April; sometimes as early as the 13th, as was the case this year (1878); the first are generally reported from the cliffs at St. Martin's, near Moulin Huet, the first land they would make on their arrival from Brittany." Very soon after their arrival, however, they spread over the whole Island of Guernsey, as well as all the other neighbouring

islands, in all of which they are equally plentiful; they seem to cross from one to the other without much considering four or five miles of sea, or being the least particular as to taking the shortest passage across from island to island. As usual, there were a great number of Cuckoos in the Vale whilst I was there this summer (1878); but I was unfortunate in not finding eggs, and in not seeing any of the foster-parents feeding their over-grown *protigis*: this was rather surprising, as there were so many Cuckoos about, and many must have been hatched and out of the nest long before we left at the end of July. I should think, however, Tree and Meadow Pipits, Skylarks and Stonechats, from their numbers and the numbers of their nests, must be the foster-parents most usually selected; other favourites, such as Wagtails, Hedgesparrows, and Robins, being comparatively scarce in that part of the Island, and Wheaters, which were numerous, had their nests too far under large stones to give the Cuckoo an opportunity of depositing her eggs there. I should have been very glad if I could have made a good collection of Cuckoos' eggs in the Channel Islands, and, knowing how common the bird was, I fully expected to do so, but I was disappointed, and consequently unable to throw any light on the subject of the variation in the colour of Cuckoos' eggs, as far as the Channel Islands are concerned, or how far the foster-parents had been selected with a view to their eggs being similar in colour to those of the Cuckoo about to be palmed off upon them. The only Cuckoos' eggs I saw were a few in the Museum, and in one or two other small collections: all these were very much the same, and what appears to me the usual type of Cuckoo's egg, a dull greyish ground much spotted with brown, and a few small black marks much like many eggs of the Tree or Meadow Pipit. It is hardly the place here to discuss the question how far Cuckoos select the nest of the birds whose eggs are similar to their own, to deposit their eggs in, or whether a Cuckoo hatched and reared by one foster-parent would be likely to select the nest of the same species to deposit its own eggs in; the whole matter has been very fully discussed in several publications, both English and German; and Mr. Dresser has given a very full *resumi* of the various arguments in his 'Birds of Europe'; and whilst fully admitting the great variation in the colour of the Cuckoos' eggs, he does not seem to think that any particular care is taken by the parent Cuckoo to select foster-parents whose eggs are similar in colour to its own; and the instances cited seem to bear out this opinion, with which, as far as my small experience goes, I quite agree.

Whilst on the subject of Cuckoos I may mention, for the information of such of my Guernsey readers who are not ornithologists, and therefore not well acquainted with the fact, the peculiar state of plumage in which the female Cuckoo occasionally returns northward in her second summer; I mean the dull reddish plumage barred with brown, extremely like that of the female Kestrel: in this plumage she occasionally returns in her second year and breeds; but when this is changed for the more general plumage I am unable to state for certain, but probably after the second autumnal moult. The changes of plumage in the Cuckoo, however, appear to be rather irregular, as I have one killed in June nearly in the normal plumage, but with many of the old feathers left, which have a very Kestrel-like appearance, being redder than the ordinary plumage of the young bird; some of the tail-feathers, however, have more the appearance of the ordinary tail-feathers of the young Cuckoo soon after the tail has reached its full growth: the moult in this bird must have been very irregular, as it was not completed in June, when, as a rule, it would have been in full plumage, unless, as may possibly be the case, this bird was the produce of a second laying during the southern migration, and consequently, instead of a year, be only about six months old. This, however, is not a very common state of plumage; but it is by no means uncommon to find a Cuckoo in May or June with a good deal of rusty reddish barred with brown, forming a sort of collar on the breast. I merely mention these rather abnormal changes of plumage, as they may be interesting to any of my Guernsey readers into whose hands a Cuckoo may fall in a state of change and prove a puzzle as to its identity. The Cuckoo departs from the Channel Islands much about the same time that it does from England on its southern migration in August or September. Occasionally, however, this southern migration during the winter seems to be doubted, as a clerical friend of mine once told me that a brother clergyman, a well educated and even a learned man, told him, when talking about Cuckoos and what became of them in winter, that "it was a mistake to suppose they migrated, but that they all turned into Sparrow-hawks in the winter." As my friend said, could any one believe this of a well-educated man in the nineteenth century?

The Cuckoo is mentioned in Professor Ansted's list, but only marked as occurring in Guernsey and Sark. There are three specimens, one adult and two young, in the Museum, as well as some very ordinary eggs.

85. KINGFISHER. *Alcedo ispida*, Linnaeus. French, "Martin Pecheur."—The Kingfisher is by no means uncommon, is generally spread over the Islands, and is resident and breeds at all events in Guernsey, if not in the other Islands also. It is generally to be seen amongst the wild rocks which surround L'Ancresse Common, where it feeds on the small fish left in the clear pools formed amongst the rocks by the receding tide; it is also by no means uncommon amongst the more sheltered bays in the high rocky part of the Island; it is also to be found about the small ponds in various gardens. About those in Candie Garden I have frequently seen Kingfishers, and they breed about the large ponds in the Vale in Mr. De Putron's grounds; they also occasionally visit the wild rocky islets to the northward of Herm, even as far as the Amfrocques, the farthest out of the lot. As well as about the Vale ponds, the Kingfisher breeds in holes in the rocks all round the Island. I have not myself seen it in Alderney, but Captain Hubbach writes me word he saw one there about Christmas, 1862. I think its numbers are slightly increased in the autumn by migrants, as I have certainly seen more specimens in Mr. Couch's shop at that time of year than at any other; this may perhaps, however, be accounted for, at all events partially, by its being protected by the Sea Bird Act during the summer and in early autumn, where the 'Martin pjcheur' appears as one of the "Oiseaux de Mer."

It is included in Professor Ansted's list, and only marked as occurring in Guernsey and Sark. There are three specimens now in the Museum.

86. NIGHTJAR. *Caprimulgus enropaeus*, Linnaeus. French, "Engoulevent ordinaire."—The Nightjar is a regular autumnal visitor, a few perhaps arriving in the spring and remaining to breed, but by far the greater number only making their appearance on their southward migration in the autumn. The Nightjar occasionally remains very late in the Islands, as Miss Carey records one in the 'Zoologist' for 1872 as occurring on the 16th of October; and I have one killed as late as the 12th of November: this bird had its stomach crammed with black beetles, not our common domestic nuisances, but small winged black beetles: these dates are later than the Nightjar usually remains in England, though Yarrell notices one in Devon as late as the 6th of November, and one in Cornwall on the 27th of November. Colonel Irby, on the faith of Fabier, says the Nightjars cross the Straits of Gibraltar on their southward journey from September to November; so these late stayers in Cornwall and Guernsey have not much time to complete their journey if they intend going as far south as the coast of Africa; perhaps, however the Guernsey ones have no such intention, as Mr. Gallienne, in his remarks published with Professor Ansted's list, says "The Nightjar breeds here, and I have obtained it summer and winter." Mr. MacCulloch tells me the Goatsucker is looked upon by the Guernsey people as a bird of ill-omen and a companion of witches in their akrial rambles. The bird-stuffer in Alderney had some wings of Nightjars nailed up behind his door which had been shot in that Island by himself.

Professor Ansted includes the Nightjar in his list, but only marks it as occurring in Guernsey and Sark. There are two specimens, a male and female, in the Museum, but no date as to time of their occurrence.

87. SWIFT. *Cypselus apus*, Linnaeus. French, "Martinet de Muraille."—The Swift is a tolerably numerous summer visitant to all the Islands, but I think most numerous in Sark, where hundreds of these birds may be seen flying about the Coupie, amongst the rocks of which place and Little Sark they breed in considerable numbers. Mr. MacCulloch and Mr. Gallienne appear to think the Swift rare in

Guernsey, as Mr Gallienne says in his remarks on Professor Ansted's list, "The swift appears here (Guernsey) in very small numbers, but is abundant in Sark;" and Mr. MacCulloch writes me word, "I consider the Swift very rare in Guernsey." I certainly cannot quite agree with this, as I have found them by no means uncommon, though certainly rather locally distributed in Guernsey. One afternoon this summer (1878) Mr. Howard Saunders and I counted forty within sight at one time about the Gull Cliff, near the old deserted house now known as Victor Hugo's house, as he has immortalised it by describing it in his 'Travailleurs de la Mer.' The Swifts use this and two similar houses not very far off for breeding purposes, a good many nesting in them, and others, as in Sark, amongst the cliffs. Young Le Cheminant had a few Swifts' eggs in his small collection, probably taken from this very house, as the Swift is certainly, as Mr. MacCulloch says, rare in other parts of Guernsey. In Alderney the Swift is tolerably common, and a good many pairs were breeding about Scott's Hotel when I was there this year (1878). Probably a good many Swifts visit the Islands, especially Alderney, for a short time on migration, principally in the autumn, as once when I was crossing from Weymouth to Guernsey, on the 18th of August, I saw a large flock of Swifts just starting on their migratory flight; they were plodding steadily on against a stormy southerly breeze, spread out like a line of skirmishers, not very high, but at a good distance apart; there was none of the wild dashing about and screeching which one usually connects with the flight of the Swift, but a steady business-like flight; they went a little to the eastward of our course in the steamer, and this would have brought them to land in Alderney or Cape la Hague.

Professor Ansted included the Swift in his list, but oddly enough, considering the remark of Mr. Gallienne above quoted, marks it as only occurring in Guernsey. There is no specimen at present in the Museum.

88. SWALLOW, *Hirundo rustica*, Linnaeus. French, "Hirondelle de Cheminie."—According to Mitivier's 'Dictionary,' "Aronde" is the local Guernsey-French name of the Swallow, which is a common

summer visitant to all the Islands, and very generally distributed over the whole of them, and not having particular favourite habitations as the Martin has. It arrives and departs much about the same time that it does in England, except that I do not remember ever to have seen any laggers quite so late as some of those in England. A few migratory flocks probably rest for a short time in the Islands before continuing their journey north or south, as the case may be; the earliest arrivals and the latest laggers belong to such migratory flocks, the regular summer residents probably not arriving quite so soon, and departing a little before those that pay a passing visit; consequently the number of residents does not appear at any time to be materially increased by such wandering flocks.

Professor Ansted includes the Swallow in his list, but only marks it as occurring in Guernsey and Sark. There is no specimen of any of the Hirundines in the Museum.

89. MARTIN. *Chelidon urbica*, Linnaeus. French, "Hirondelle de fenjtre."—The House Martin is much more local than the Swallow, but still a numerous summer visitant, like the Swallow, arriving and departing about the same time that it does in England. It is spread over all the Islands, but confined to certain spots in each; in Guernsey the outskirts of the town about Candie Road, and the rocks in Fermain and Petit Bo Bay, seem very favourite nesting-places. In Alderney there were a great many nests about Scott's Hotel and a few more in the town, but I did not see any about the cliffs as at Fermain and Petit Bo in Guernsey.

Professor Ansted includes it in his list, but only marks it as occurring in Guernsey and Sark.

90. SAND MARTIN. *Cotyle riparia*, Linnaeus. French, "Hirondelle de rivage."—When I first made out my list of Guernsey birds I had omitted the Sand Martin altogether, as I had never seen it in the Islands, but Mr. MacCulloch wrote to me to say, "Amongst the swallows you have not noticed the Sand Martin, which is our earli-

est visitant in this family and by no means uncommon." In consequence of this note, as soon as I got to the Island this year (1878), in June, I went everywhere I could think likely to look for Sand Martins, but nowhere could I find that the Sand Martins had taken possession of a breeding-station. Knowing from my own experience here that Sand Martins are fond of digging their nest-holes in the heads of quarries, (I had quite forty nest-holes in my quarry this year, and forty pairs of Sand Martins inhabiting them), I kept a bright look-out in all the stone-quarries in the Vale, and they are very numerous, but I did not see a single Sand Martin's hole or a single pair of birds anywhere; and it appeared to me that the sandy earth forming the head was not deep enough before reaching the granite to admit of the Sand Martins making their holes; and they do not appear to me to have fixed upon any other sort of breeding place in the Island; neither could Mr. MacCulloch point one out to me; so I suppose we must consider the Sand Martin as only a spring visitant to this Island, not remaining to breed. The same seems to me to be the case in Alderney, as Captain Hubbach writes to tell me he "saw some Sand Martins about the quarry here (in Alderney), for two or three days at the beginning of April, but cannot say whether they remained here to breed or not." I suppose they continued their journey, as I did not see any when there in June; I have not seen any in Sark or either of the other small Islands.

Professor Ansted includes the Sand Martin in his list, and marks it as occurring in Guernsey and Sark.

91. WOOD PIGEON. *Columba palumbus*, Linnaeus. French, "Colombe ramier." — The Wood Pigeon is resident and breeds in several places in Guernsey; but fortunately for the Guernsey Farmers, who may congratulate themselves on the fact, the Wood Pigeons do not breed in very great numbers. I may mention the trees in the New Ground, Candie Garden, the Vallon and Woodlands, as places where Wood Pigeons occasionally breed. No doubt the number of Wood Pigeons is occasionally increased by migratory, or rather perhaps wandering, flocks, as Mr. Couch, in a note to the 'Zoologist,' dated October the 21st, 1871, says, "On Tuesday a great num-

ber of Wood Pigeons rested and several were shot." Mr. MacCulloch also writes me, "The Wood Pigeon occasionally arrives in large numbers. A few years ago I heard great complaints of the damage they were doing to the peas;" [15] but luckily for the farmers these wandering flocks do not stay long, or there would be but little peas, beans, or grain left in the Islands; and the Wood Pigeons would be more destructive to the crops in Guernsey than in England, as there are not many acorns or Beech masts on which they could feed; consequently they would live almost entirely on the farmer; and to show the damage they would be capable of doing in this case, I may say that in the crops of two that I examined some time ago—not killed in Guernsey however—I found, in the first, thirty seven beech-masts in the crop, and eight others in the gizzard, sufficiently whole to be counted; and in the crop of the other the astonishing number of seventy-seven beech-masts and one large acorn; the gizzard of this one I did not examine. I only mention this to show the damage a few Wood Pigeons would do supposing they were restricted almost entirely to agricultural produce for their food, as they would be in Guernsey if they lived there in any great numbers.

The Wood Pigeon is mentioned by Professor Ansted and marked as only occurring in Guernsey, and probably as far as breeding is concerned this is right (of course with the exception of Jersey); but wandering flocks probably occasionally visit Alderney as well. There is no specimen in the Museum.

92. ROCK DOVE. *Columba livia*, Linnaeus. French, "Colombe biset."—I have never seen the Rock Dove in any of the Islands, though there are many places in all of them that would suit its habits well; and Mr. MacCulloch writes to me to say, "I have heard that in times past the Rock Pigeon used to breed in large numbers in the caves around Sark"; but this certainly is not the case at present. Captain Hubbach also writes to me from Alderney, "There were some Rock Doves here in the winters of 1862 and 1863; I shot two or three of them then." Probably a few yet remain in both Alderney and Sark, though they certainly are not at all numerous in either island.

Professor Ansted includes the Rock Dove in his list, and marks it as occurring in Guernsey and Sark. There is no specimen in the Museum. Professor Ansted also includes the Stock Dove, *Columba aenas*, Linnaeus, in his list as occurring in Guernsey and Sark; but I think he must have done so on insufficient evidence, as I have never seen it and not been able to gain any information about it; neither does Mr. Gallienne say anything about it in his notes appended to the list; so on the whole I think it better to omit it in my list; but as it may occur at any time, especially as it is certainly increasing considerably in numbers in the West of England, I may mention that it may be immediately distinguished from the Rock Dove by the absence of the white rump, that part being nearly the same colour as the back in the Stock Dove, and from the Wood Pigeon, *Columba palumbus*, by its smaller size and the entire absence of white on the wing. It is perhaps more necessary to point out this difference, as the Stock Dove frequently goes by the name of the Wood Pigeon; indeed Dresser has adopted this name for it, the Wood Pigeon being called the Ring Dove, as is very frequently the case.

93. TURTLE DOVE. *Turtur vulgaris*, Eyton. French, "Colombe tourterelle." — The Turtle Dove is a regular, but probably never very numerous summer visitant, arriving and departing about the same time as in England. Neither Miss Carey nor Mr. Couch ever mention it in their notes on Guernsey birds in the 'Zoologist': and Mr. MacCulloch, writing to me about the bird, does not go farther than to say "The Turtle Dove has, I believe, been known to breed here." In June, 1866, however, I shot one in very wild weather, flying across the bay at Vazon Bay; so wild was the weather with drifting fog and rain that I did not know what I had till I picked it up; in fact, when I shot it I thought it was some wader, flying through the fog towards me. This summer (1878) I saw two at Mr. Jago's which had been shot at Herm in May, just before I came; and in June I saw one or two more about in Guernsey. The pair shot in Herm would probably have bred in that island if they had been left unmolested.

Professor Ansted mentions it in his list, but only as occurring in Guernsey, and there is one specimen in the Museum.

94. QUAIL. *Coturnix communis*, Bonnaterre. French, "Caille."—I have never seen the Quail in the Islands myself, and it cannot be considered more than an occasional straggler; there can be no doubt, however, that it sometimes remains to breed, as there are some eggs in the Museum which I have reason to believe are Guernsey taken, and Mr. MacCulloch writes me word that "Quails certainly visit us occasionally, and I remember having seen their eggs in my youth"; and Mrs. Jago (late Miss Cumber), who was herself a bird-stuffer in Guernsey a good many years ago, told me she had had two Quails through her hands during the time she had been stuffing; but evidently she had not had very many, nor did she think them very common, as she did not know what they were when they were brought to her, and she was some time before she found anyone to tell her. The Quail breeds occasionally, too, in Alderney, as the bird-stuffer and carpenter had some Quail's and Landrail's eggs; these he told me he had taken out of the same nest which he supposed belonged originally to the Landrail, as there were rather more Landrail's than Quail's eggs in it.

Professor Ansted includes the Quail in his list, but marks it as occurring only in Guernsey. There is a specimen in the Museum, and, as I said before, several eggs.

95. WATER RAIL. *Rallus aquations*, Linnaeus. French, "Rble d'eau."—The Water Rail is not very common in Guernsey, but a few occur about the Braye Pond, and in other places suited to them; and, I believe, occasionally remain to breed, as Mr. Jago, the bird-stuffer, told me he had seen a pair of Water Rails and four young, his dog having started them from a hedge near the Rousailleries farm; the young could scarcely fly. I saw one at the bird-stuffer's at Alderney, which had been shot in that Island; and the bird-stuffer told me they were common, and he believed they bred there, but he had no eggs. Their number, however, is, I think, rather increased in the autumn by migrants; at all events, more specimens are brought to the bird-stuffers at that time of year. I have before mentioned the incident of

the Water Rail being killed by the Merlin, recorded by Mr. Couch in the 'Zoologist' for 1875.

The Water Rail is included in Professor Ansted's list, and marked as occurring in Guernsey and Sark. There are two specimens in the Museum.

96. SPOTTED CRAKE. *Porzana maruetta*, Leach. French, "Poule d'eau marouette."—I have some doubt as to the propriety of including the Spotted Crake in my list, but, on the whole, such evidence as I have been able to collect seems in favour of its being at all events occasionally seen and shot, though its small size and shy skulking habits keep it very much from general notice. Mr. MacCulloch, however, writes to me to say the Spotted Rail has been found here; and one of Mr. De Putron's labourers described a Rail to me which he had shot in the Vale Pond in May, 1877, which, from his description, could have been nothing but a Spotted Rail.

This is all the information I have been able to glean, but Professor Ansted includes it in his list, and marks it as occurring in Guernsey. There are also two pretty good specimens in the Museum, which I have no doubt were killed in Guernsey.

97. LANDRAIL. *Crex pratensis*, Bechstein. French, "Rble des pris," "Rble de terre" ou "de Genet," "Poule d'eau de genet."—The Landrail is a common summer visitant, breeding certainly in Guernsey, Sark, and Alderney, [16] and probably in Herm, though I cannot be quite so sure about the latter Island. It seems to be rather more numerous in some years than others, as occasionally I have heard them craking in almost every field. But the last summer I was in the Islands (1878) I heard very few. The Corn Crake arrives and departs much about the same time as in England, and I have never been able to find that any stay on into the winter, or even as late as November.

It is included in Professor Ansted's list, but only marked as occurring in Guernsey and Sark. There are two specimens in the Museum.

98. MOORHEN. *Gallinula chloropus*, Linnaeus. French, "Poule d'eau ordinaire." — I have not seen the Moorhen myself in Guernsey, but Mr. Couch, writing to me in December, 1876, told me that Mr. De Putron informed him that Coots, Waterhens, and Little Grebes bred that year in the Braye Pond; and Mr. De Putron, to whom I wrote on the subject, said the information I had received was perfectly correct. I see no reason to doubt the fact of the Moorhen occasionally breeding in Mr. De Putron's pond, and perhaps in other places in the Island, especially the Grand Mare. But I do not believe they breed regularly in either place; they certainly did not in this last summer (1878), or I must have seen or heard them. As far as Mr. De Putron's pond is concerned, I could not have helped hearing their loud call or alarm note had only one pair been breeding there; I have, however, a young bird of the year, killed in Guernsey in November, 1878.

Professor Ansted includes it in his list, and marks it as only occurring in Guernsey. There are two specimens in the Museum, probably both Guernsey killed.

99. COMMON COOT. *Fulica atra*, Linnaeus. French, "Foulque," "Foulque macroule." — In spite of Mr. De Putron's statement that the Coot bred in the Braye Pond in the summer of 1876, I can scarcely look upon it in the light of anything but an occasional and never numerous autumnal visitant; and its breeding in the Braye Pond that year must have been quite exceptional. In the autumn it occurs both in the Braye Pond and on the coast in the more sheltered parts. I have the skin of one killed in the Braye Pond in November, 1876, which might have been one of those bred there that year.

Professor Ansted includes the Coot in his list, but only marks it as occurring in Guernsey. There is no specimen in the Museum.

100. LITTLE BUSTARD. *Otis tetrax*, Linnaeus. French, "Outarde canepetihre," "Poule de Carthage."—The Little Bustard can only be considered a very rare occasional visitor to the Channel Islands, and very few instances of its occurrence have come under my notice. The first was mentioned to me by Mr. MacCulloch, who wrote me word that a Little Bustard was killed in Guernsey in 1865, but unfortunately he gives no information as to the time of the year. Another was shot by a farmer in Guernsey early in March, 1866, and was recorded by myself in the 'Zoologist' for that year. Mr. Couch also recorded one in the 'Zoologist' for 1875, "as having been shot at the back of St. Andrew's (very near the place where one was shot fifteen years ago) on the 20th of November, 1874." This bird is now in the possession of Mr. Le Mottee, at whose house I saw it, and was informed that it had been shot at a place called the Eperons, in the parish of St. Andrew's, on the date above mentioned. These are all the instances of the occurrence of the Little Bustard in the Channel Islands that I have been able to gain any intelligence of, but they are sufficient to show that although by no means a common visitor, it does occasionally occur on both spring and autumn migration.

It is not included in Professor Ansted's list. There is, however, a specimen in the Museum, which I was told, when I saw it in 1866, had been killed the previous year, but there is no date of the month, and I should think, from the state of plumage, it was an autumn-killed specimen: it is still in the Museum, as I saw it there again this year, 1878. This is probably the bird mentioned by Mr. MacCulloch as killed in 1865, and also very likely the one spoken of by Mr. Couch, in 1875, as having been killed in St. Andrew's fifteen years ago; but there seems to have been some mistake as to Mr. Couch's date for this one, as, had it been killed so long ago as 1860, it would in all probability have been included in Professor Ansted's list, and mentioned by Mr. Gallienne in his remarks on some of the birds included in the list.

101. THICK-KNEE. *Oedicnemus scolopax*, S.G. Gmelin. French, "Oedicneme criard," "Poule d'Aurigny." [17] —The Thick-knee, Stone Curlew, or Norfolk Plover, as it is called, though only an occasional

visitant, is much more common than the Little Bustard; indeed, Mr. MacCulloch says that "it is by no means uncommon in winter. The French call it 'Poule d'Aurigny,' from which one might suppose it was more common in this neighbourhood than elsewhere." Miss C.B. Carey records one in the 'Zoologist' as killed in November, and Mr. Couch another as having been shot on the 31st December. I have also seen one or two hanging up in the market, and others at Mr. Couch's, late in November; and one is recorded in the 'Guernsey Mail and Telegraph' as having been shot by Mr. De Putron, of the Catel, on the 3rd January, 1879. From these dates, as well as from Mr. MacCulloch's remark that it is not uncommon in the winter, it would appear that—as in the Land's End district in Cornwall—the Thick-knee reverses the usual time of its visits to the British Islands, being a winter instead of a summer visitant; and probably for the same reason, namely, that the latitude of the Channel Islands, like that of Cornwall, is about the same as that of its most northern winter range on the Continent.

Professor Ansted includes it in his list, but only marks it as occurring in Guernsey. There is one specimen in the Museum.

102. PEEWIT. *Vanellus vulgaris*, Bechstein. French, "Vanneau huppi."—The Peewit is a common and rather numerous autumn and winter visitant to all the Islands, though I have never seen it in such large flocks as in some parts of England, especially in Somerset. Those that do come to the Islands appear to take very good care of themselves, for I have always found them very difficult to get a shot at, and very few make their appearance in the market. Though generally a winter visitant, I have seen occasional stragglers in summer. On the 9th July this year (1878), for instance, I saw one fly by me in L'Ancresse Bay; this was either a young bird, or, if an adult, was not in breeding plumage, as I could clearly see that the throat was white—- not black, as in the adult in breeding plumage. A few days afterwards, July 19th, another—or, perhaps, the same—was shot by some quarry-men on the common; this was certainly a young bird of the year, and I had a good opportunity of looking at it. In spite of occasional stragglers of this sort making their appearance in the

summer, I have never been able to find that the Peewit breeds on any of the Islands; but, by the 9th of July, stragglers, both old and young, might easily come from the opposite coast of Dorsetshire, where a good many breed, or from the north of France.

Professor Ansted includes the Peewit in his list, but only marks it as occurring in Guernsey. There is no specimen in the Museum at present.

103. GREY PLOVER. *Squatarola helvetica*, Linnaeus. French, "Vanneau pluvier."—The Grey Plover is a regular but by no means numerous visitor to the coast of all the Islands during the winter months, but I have never found it in flocks like the Golden Plover. A few fall victims to the numerous gunners who frequent the shores during the autumn and winter, and consequently it occasionally makes its appearance in the market, where I believe it often passes for a Golden Plover, especially in the case of young birds on their first arrival in November; but for the sake of the unknowing in such matters, I may say that they need never be deceived, as the Grey Plover has a hind toe, and also has the axillary plume or the longish feathers under the wing black, while the Golden Plover has no hind toe and the axillary plume white: a little attention to these distinctions, which hold good at all ages and in all plumages, may occasionally save a certain amount of disappointment at dinner time, as the Grey Plover is apt to taste muddy and fishy, and is by no means so good as the Golden Plover.

It is included in Professor Ansted's list, but only marked as occurring in Guernsey. There are two specimens in the Museum, both in winter plumage. Indeed, I do not know that it even remains long enough in the Channel Islands to assume, even partially, the blackbreast of the breeding plumage, as it so often does in England.

104. GOLDEN PLOVER. *Charadrius pluvialis*, Linnaeus. French, "Pluvier dore."—A common winter visitant to all the Islands, arriving about the end of October or beginning of November, and re-

maining till the spring, sometimes till they have nearly assumed the black breast of the breeding-season; but I do not know that the Golden Plover ever breeds in the Islands, at all events in the present day.

Professor Ansted includes the Golden Plover in his list, and marks it as occurring in Guernsey and Sark. There is one specimen in the Museum, probably killed rather late in the spring, as it is assuming the black breast.

105. DOTTEREL. *Eudromias morinellus*, Linnaeus. French, "Pluvier guignard."—The common Dotterel is a rare occasional visitor to the Channel Islands, occurring, however, on both the spring and autumn migration, as Mr. MacCulloch says he has a note of a Dotterel killed in May, 1849; he does not say in which of the Islands, but probably in Guernsey; and I have a skin of one, a fine full-plumaged bird, according to Mr. Couch, who forwarded me the skin, a female by dissection, killed in Herm on the 26th of April, 1877. Another skin I have is that of a young bird of the year, killed in the autumn, I should think early in the autumn—August or September; and the Rev. A. Morrks, who kindly gave me this last one, has also a skin of one killed at the same time; both of these were Guernsey killed.

The Dotterel is included in Professor Ansted's list, and by him marked as having occurred in Guernsey and Sark. I should think Alderney a more likely place for the bird to have occurred than Sark, but I have not been able to gain any information about its occurrence there; neither the carpenter bird-stuffer nor his sporting friend had a skin or any part of the bird. There is no specimen now in the Museum.

106. RING DOTTEREL. *Fgialitis hiaticula*, Linnaeus. French, "Grand pluvier ` collier," "Pluvier ` collier."—The Ring Dotterel is very common in all the Islands in places suited to it. Some remain throughout the summer, and a few of these, but certainly very few, may breed in the Islands; the great majority, however, of those that

frequent the coast in the winter are migrants, arriving in the autumn and departing again in the spring. Some, however, appear to arrive very early, and cannot have bred very far off, perhaps on the neighbouring coast of France or Dorset. I have the following note on the subject in the 'Zoologist' for 1866, which gives the time of their arrival pretty correctly. During the first two or three weeks after my arrival—that was on the 21st of June, 1866—I found Ring Dotterels excessively scarce even on parts of the coast, where, on other visits later in the year, I had found them very numerous. Towards the middle of July, however, they began to frequent their usual haunts in small parties of six or seven, most probably the old birds with their young. These parties increased in number to twenty or thirty, and before my departure, on the last day of July, they mustered quite as thickly as I had ever seen them before. On another summer visit to Guernsey, from the 3rd to the 19th of June, 1876, I did not see any Ring Dotterel at all, though at the time Kentish Plover were common in most of the bays in the low parts of the Island. The Ring Dotterel must therefore have selected some breeding-place separate from the Kentish Plover, probably not very far off; but I do not believe it breeds at all commonly in the Islands. This agrees very much with what I saw of the Ring Dotterel this year (1878); there were a few in L'Ancresse and one or two other bays, but none in Grand Havre, close to which I was living, and I very much doubt if any of those I saw were breeding. Neither Colonel l'Estrange nor I found any eggs, though we searched hard for them both in '76 and '78; neither did we find any eggs either in Herm or Alderney.

Professor Ansted includes the Ring Dotterel in his list, but marks it as only occurring in Guernsey. There is a specimen in the Museum.

107. KENTISH PLOVER. *Fgialitis cantianus*, Latham. French, "Pluvier ` collier interrompu." I have always looked upon the Kentish Plover as only a summer visitant to the Islands, never having seen it in any of my visits in October and November; but Mr. Harvie Brown mentions ('Zoologist' for 1869) seeing some of these birds in January, at Herm, feeding with the Ring Dotterel, but he says they

always separated when they rose to fly. If he is not mistaken, which my own experience inclines me to think he was, we must look upon the Kentish Plover as partially resident in the Islands, the greater number, however, departing in the autumn. Until this summer (1878) I have been unsuccessful in finding the eggs of the Kentish Plover, though I have had many hard searches for them; and they are very difficult to find, unless the bird is actually seen to run from the nest, or rather from the eggs, for, as a rule, nest there is none, the eggs being only placed on the sand, with which they get half buried, when they may easily be mistaken for a small bit of speckled granite and passed by. In the summer of 1866, a friend and myself had a long search for the eggs of a pair we saw and were certain had eggs, as they practised all the usual devices to decoy us from them, till my friend, actually thinking one of the birds to be badly wounded, set his dog at it; after this all chance was over: this was in a small sandy bay, called Port Soif, near the Grand Rocques Barracks. I mention this as I am certain these birds had eggs or young somewhere close to us, and this was the farthest point towards Vazon Bay from the Vale I found them breeding. The sandy shores of Grand Havre and L'Ancresse Bay seemed to be their head breeding-quarters in Guernsey. Though I only found one set of eggs in Grand Havre, I am sure there were three or four pairs of birds breeding there; the two eggs I found were lying with their thick ends just touching each other and half buried in sand; there was no nest whatever, not even the sand hollowed out; they were in quite a bare place, just, and only just, above the high-water line of seaweed. I should not have found these if it had not been for the tracks of the birds immediately round them. In L'Ancresse Bay I was not equally fortunate, but there were quite as many pairs of birds breeding there. In Herm the shell-beach seems to be their head breeding-quarters, and there Mr. Howard Saunders, Colonel l'Estrange and myself found several sets of eggs, generally three in number, but in one or two instances four: these were probably hard-sat; in one instance, with four eggs, the eggs were nearly upright in the sand, the small end being buried, and the thick end just showing above the sand. In no instance in which I saw the eggs was there the slightest attempt at a nest; but Colonel l'Estrange told me that in one instance, in which he had found some eggs a day or two before I got to Guernsey, quite the end of May, he found there was a slight attempt at a nest, a few bents of the rough herbage which grew in the sand just above high-water mark having been collected and the nest lined with them. I have not found any eggs in Alderney, but I have no doubt they breed in some of the sandy bays to the north of the Island occasionally, if not always, as I have seen them there in the breeding-season, both in 1876 and in 1866. This summer (1878) I was so short a time in that Island that I had not time to search the most likely places, but Captain Hubbach wrote me—"I do not think the Kentish Plover remained here to breed this year, although I saw some about in April."

Professor Ansted includes the Kentish Plover in his list, but only marks it as occurring in Guernsey. There is one specimen, a male, in the Museum.

108. TURNSTONE. *Strepsilas interpres*, Linnaeus. French, "Tourne pierre," "Tourne pierre a collier." The cosmopolitan Turnstone is resident in the Channel Islands; throughout the year its numbers, however, are much increased in the autumn by migrants, many of which remain throughout the winter, leaving the Islands for their breeding-stations in the spring. Some of those that remain throughout the summer I have no doubt breed in the Islands, as I have seen the old birds about with their young and shot one in July; and on the 8th of June, 1876, I saw a pair in full breeding plumage in L'Ancresse Bay; I saw them again about the same place on the 16th: these birds were evidently paired, and I believe had eggs or young on a small rocky island about two or three hundred yards from the land, but there was no boat about, and so I could not get over to look for the eggs. Col. l'Estrange obtained some eggs on one of the rocky islands to the north of Herm, which certainly were not Tern's eggs as he supposed, and I believe them to have been Turnstone's; unluckily he did not take the eggs himself, but the boatman who was with him took them, so he did not see the bird go off the nest. This last summer (1878) I was in hopes of being more successful either in Guernsey itself or in Herm, or the rocks near there, but I did not see a single Turnstone alive the whole time I was in Guernsey. I think it very likely, however, I should have been successful in Herm, as I

visited it several times both by myself and with Col. l'Estrange and Mr. Howard Saunders; our first visit was on June the 21st, when we did not see a single Turnstone; but this was afterwards accounted for, as on a visit to Jago, the bird-stuffer, a short time afterwards, I found him skinning a splendid pair of Turnstones which had been shot in Herm a few days before our visit on the 17th or 18th of June; the female had eggs ready for extrusion; I need not say I did not exactly bless the person who, in defiance of the Guernsey Sea Birds Act, had shot this pair of Turnstones, as had they been left I have no doubt we should have seen them, and probably found the eggs, and quite settled the question of the Turnstone's breeding there. I have long been very sceptical on this subject, but now I have very little doubt, as I think, seeing the birds about, paired, in Guernsey in June and the pair shot in Herm, the female with eggs in June, pretty well removes any doubt as to the Turnstone breeding in the Islands, and I do not see why it should not, as it breeds quite as far south in the Azores, and almost certainly in the Canaries. [18] Mr. Rodd, however, tells me he does not believe in its breeding in the Scilly Islands, though it is seen about there throughout the year, as it is in the Channel Islands. Mr. Gallienne, in his remarks on Professor Ansted's list, merely says, "The Turnstone is found about the neighbourhood of Herm throughout the year." It occurs also in Alderney in the autumn, but I have not seen it there in the breeding-season.

Professor Ansted includes it in his list, but only marks it as occurring in Guernsey. There are a male and female, in breeding plumage, in the Museum, and also one in winter plumage.

109. OYSTERCATCHER, *Haematopus ostralegus*, Linnaeus. French, "Hi{trier pie."—The Guernsey Bird Act includes these birds under the name 'Piesmarans,' which is the name given to the Oystercatcher by all the French-speaking fishermen and boatmen, and which I suppose must be looked upon only as the local name, though I have no doubt it is the common name also on the neighbouring coast of Normandy and Brittany. The Oystercatcher is resident all the year, and breeds in all the Islands; I think, however, its numbers are considerably increased in the autumn by migratory arrivals; certainly

the numbers actually breeding in the Islands are not sufficient to account for the immense flocks one sees about in October and November. There seem, however, to be considerable numbers remaining in flocks throughout the summer, without apparently the slightest intention of separating for breeding purposes, as I have often counted as many as forty or fifty together in June and July. The Oystercatcher breeds in Guernsey itself about the cliffs. Mr. Howard Saunders, Colonel l'Estrange and myself found one very curiously placed nest of the Oystercatcher on the ridge of a hog-backed rock at the bottom of the cliff, near the south end of the Island; it was not much above high-water mark, and quite within reach of heavy spray when there was any sea on: we could distinctly see the eggs when looking down from the cliffs on them, and the two old birds were walking about the ridge of rock as if dancing on the tight-rope; how they kept their eggs in place on that narrow ridge, exposed as it was to wind and sea, was a marvel. The Oystercatcher breeds also in both the small Islands, Jethou and Herm, on almost all the rocky islands to the north of Herm, in Sark and Alderney, and on Burhou, near Alderney, where I found one clutch of three of the most richly marked Oystercatcher's eggs I ever saw: these, as well as another clutch, also of three eggs, were placed on rather curious nests; they were on the smooth rock, but in both cases the birds had collected a number of small stones and made a complete pavement of them, on which they placed their eggs; there was no protection, however, to prevent the eggs from rolling off. Both in Burhou as well as on the Amfroques and other rocks to the north of Herm, the eggs of the Oystercatchers, as well as of the other sea-birds breeding there, had been ruthlessly robbed by fishermen and others, who occasionally visit these wild rocks and carry off everything in the shape of an egg, without paying any respect to the Bird Act, which professes to protect the eggs as well as the birds.

Professor Ansted includes the Oystercatcher in his list, but only marks it as occurring in Guernsey and Sark. There is an Oystercatcher and also a few of the eggs in the Museum.

110. CURLEW. *Numenins arquata*, Linnaeus. French, "Courlis," "Grand courlis cendri."—A good many Curlews are to be found in the Islands throughout the year, but I do not believe any of them breed there; I have seen them in Guernsey, Jethou, Herm and Alderney, all through the summer, but always in flocks on the mud and seaweed below high-water mark, whenever they can be there, searching for food, and quite as wild and wary as in the winter. I have never seen them paired, or in any place the least likely for them to be breeding. I know Mr. Gallienne, in his remarks to Professor Ansted's list, says, "Although I have never heard of the eggs of either the Curlew or Whimbrel being found, I am satisfied they breed here (I think at Herm), as they stay with us throughout the year." I cannot from my observation agree with this supposition of the Curlew breeding in the Islands; nor can I agree with the statement made by a writer in 'Cassel's Magazine' for June or July, 1878, that he found a young Curlew in the down on one of the Islands near Jethou, probably from the description 'La Fauconnihre.' The writer of this paper in 'Cassel's Magazine' was evidently no ornithologist, and must, I think, have mistaken a young Oystercatcher, of which several pairs were breeding there at the time, for a young Curlew; his description of the cry of the old birds as they flew round was much more like that of the Oystercatcher than the Curlew. All of the boatmen also, with whom I have been about at various times, agree that the Curlews do not breed in the Islands, though they are quite aware that they remain throughout the year, and as many of them, in spite of the Guernsey Bird Act, are great robbers of the eggs of the Gulls, Puffins, and Oystercatchers, all of which they know well, they would hardly miss such a fine mouthful as the egg of the Curlew if it was to be found. No doubt the number of Curlews is largely increased in the autumn by migratory visitors, which remain throughout the winter and depart again in the spring: though numerous during autumn and winter, they are very wild and wary, and, as everywhere else where I have had any experience of Curlews at that time of year, very difficult to get a shot at; consequently very few find their way into the market.

The Curlew is mentioned in Professor Ansted's list, but only marked as occurring in Guernsey and Sark. There are two specimens in the Museum.

111. WHIMBREL. *Numenius phaeopus*, Linnaeus. French, "Courlis corlieu." — A good many Whimbrel visit all the Islands during the spring migration, and a few may stay some little time into the summer, as I have seen them as late as June, but, as far as I have been able to make out, none breed there; a few also may make their appearance on the autumn migration, but very few in comparison with those which appear in the spring, and I have never seen any there at that time. Purdy, one of the Guernsey boatmen, who is pretty well up in the sea and shore birds, told me the Whimbrel occurred commonly in May, but not on the autumn migration. He added that it was known there as the "May-bird," and was very good to eat, and much easier to shoot than a Curlew, in which he is quite right.

Professor Ansted includes the Whimbrel in his list, and marks it only as occurring in Guernsey and Sark. There are two specimens in the Museum.

112. REDSHANK. *Totanus calidris*, Linnaeus. French, "Chevalier gambette." — An occasional but never numerous visitant to all the Islands, on both spring and autumn migrations; none appear to remain through the summer. I have, however, a Redshank in full breeding plumage, killed in Guernsey as late as the 23rd of April.

Professor Ansted includes the Redshank in his list, but only marks it as occurring in Guernsey. There are two specimens in the Museum.

113. GREEN SANDPIPER. *Totanus ochropus*, Linnaeus. French, "Chevalier cul blanc." — The Green Sandpiper is an irregular, very scarce (not so numerous indeed as the Redshank) visitant on the spring and autumn migration. I have seen what was probably a family party about Vazon Bay, in Guernsey, quite at the end of July, but I do not believe this bird ever breeds in the Islands: those I saw were probably the parents and young brood of an early-breeding pair, on their return from some not very distant breeding-ground.

Such parties seem only to pay the Islands a very short visit on their return from their breeding-ground; at least I have never seen a Green Sandpiper in the Islands as late as October or November; it may, however, occasionally occur in the winter, as I have a specimen from Torbay killed in December.

Professor Ansted does not include the Green Sandpiper in his list, though he does the Wood Sandpiper, giving, however, no locality for it. I have never seen this latter bird in the Islands, however; nor have I been able to find that one has ever passed through the hands of any of the local bird-stuffers, and I cannot help thinking a mistake has been made; as both birds may, however, occur, and they are something alike, I may, for the benefit of my Guernsey readers, mention that they may immediately be distinguished; the axillary plume or long feathers under the wing, in the Green Sandpiper, being black narrowly barred with white; and in the Wood Sandpiper the reverse, white with a few dark bars and markings; the tail also, in the Green Sandpiper, is much more distinctly and boldy barred with black and white. Alive and on the wing they may be immediately distinguished by the pure white rump and tail-coverts of the Green Sandpiper, which are very conspicuous, especially as the bird rises; the white on the same parts of the Wood Sandpiper is much marked with brown, and consequently never appears so conspicuously. There is one Green Sandpiper at present in the Museum, which there seems no reason to doubt is Guernsey killed.

114. COMMON SANDPIPER. *Totanus hypoleucos*, Linnaeus. French, "Chevalier guignette."—The Common Sandpiper, or Summer Snipe as it is sometimes called, is a spring and autumn visitant, but never a numerous one, sometimes, however, remaining till the summer. One of Mr. De Putron's men told me he had seen one or two about their pond all this summer (1878), and he believed they bred there; but as to this I am very sceptical; I could see nothing of the bird when I visited the pond in June and July, and I fancy the birds stayed about, as they do sometimes about my own pond here in Somerset, till late perhaps in May, and then departed to breed elsewhere. The latest occurrence I know of was one recorded by Mr.

Couch in the 'Zoologist' for 1874, as having been killed on the 3rd of October. Mr. Couch adds that this was the first specimen of the Common Sandpiper he had had since he had been in the Islands.

The Common Sandpiper is included in Professor Ansted's list, and marked as occurring in Guernsey and Sark. There is no specimen in the Museum.

115. BARTAILED GODWIT. *Limosa lapponica*, Linnaeus. French, "Barge rousse."—The Bar-tailed Godwit is a regular and sometimes rather numerous spring and autumn visitor. In May, 1876, a considerable number of these birds seem to have rested on the little Island of Herm, where the keeper shot three of them; two of these are now in my possession, and are very interesting, as though all shot at the same time—I believe on the same day—they are in various stages of plumage, the most advanced being in thorough breeding-plumage, and the other not nearly so far advanced; and the third, which I saw but have not got, was not so far advanced as either of the others. In the two which I have the change of colour in the feathers, without moult, may be seen in the most interesting manner, especially in the least advanced, as many of the feathers are still parti-coloured, the colouring matter not having spread over the whole feather; in the most advanced, however, nearly all the feathers were fully coloured with the red of the breeding-plumage. This red plumage remains till the autumn, when it is replaced, after the moult, by the more sombre and less handsome grey of the winter plumage. Though the Bar-tailed Godwit goes far north to breed, not breeding much nearer than Lapland and the north of Norway and Sweden, both old and young soon show themselves again in the Channel Islands on their return journey, as I shot a young bird of the year in Herm the last week in August. Most of the autumn arrivals, however, soon pass on to more southern winter quarters, only a few remaining very late, perhaps quite through the winter, as I have one shot in Guernsey as late as the 14th of December; this one, I need hardly say, is in full winter plumage, and of course presents a most striking difference to the one shot in Herm in May.

The Bar-tailed Godwit is included in Professor Ansted's list, but only marked as occurring in Guernsey. It is, however, as I have shown, perhaps more common in Herm, and it also occurs in Alderney. There is a series of these in the Museum in change and breeding-plumage.

The Blacktailed Godwit is also included in Professor Ansted's list, but I have never seen the bird in the Islands or been able to glean any information concerning it, and there is no specimen in the Museum.

116. GREENSHANK. *Totanus canescens*, Gmelin. French, "Chevalier gris," "Chevalier aboyeur."—The Greenshank can only be considered a rare occasional visitor. I have never shot or seen it myself in the Islands, but Miss C.B. Carey records one in the 'Zoologist' for 1872 as having been shot on the 2nd of October of that year, and brought to Mr. Couch's, at whose shop she saw it.

The Greenshank is included in Professor Ansted's list, but there is no letter to note which of the Islands it has occurred in. There is no specimen in the Museum.

117. RUFF. *Machetes pugnax,* Linnaeus. French, "Combatant," "Combatant variable."—The Ruff is an occasional but not very common autumn and winter visitor; it occurs, probably, more frequently in the autumn than the winter. Mr. MacCulloch writes me, "I have a note of a Ruff shot in October, 1871." This probably was, like all the Guernsey specimens I have seen, a young bird of the year in that state of plumage in which it leads to all sorts of mistakes, people wildly supposing it to be either a Buff-breasted or a Bartram's Sandpiper. Miss C.B. Carey records one in the 'Zoologist' for 1871 as shot in September of that year; this was a young bird of the year. Miss C.B. Carey also records two in the 'Zoologist' for 1872 as having been shot about the 13th of April in that year; these she describes as being in change of plumage but having no ruff yet; probably the change of colour in the feathers was beginning before the long feathers of the ruff began to grow; and this agrees with what I have seen of the Ruff in confinement; the change of colour in the feathers of the body begins before the ruff makes its appearance.

Professor Ansted includes the Ruff in his list, and only marks it as occurring in Guernsey. There is no specimen in the Museum at present.

118. WOODCOCK. *Scolopax rusticola*, Linnaeus. French, "Becasse ordinaire."—The Woodcock is a regular and tolerably common autumnal visitor to all the Islands, arriving and departing about the same time as in England,—none, however, remaining to breed, as is so frequently the case with us. There might be some good cock shooting in the Islands if the Woodcocks were the least preserved, but as soon as one is heard of every person in the Island who can beg, borrow, or steal a gun and some powder and shot is out long before daylight, waiting for the first shot at the unfortunate Woodcock as soon as there should be sufficient daylight. In fact, such a scramble is there for a chance at a Woodcock that a friend of mine told me he got up long before daylight one morning and went to a favourite spot to begin at; thinking to be first on the ground, he sat on a gate close by waiting for daylight; but so far from his being the first, he found, as it got light, three other people, all waiting, like himself, to begin as soon as it was light enough, each thinking he was going to be first and have it all his own way with the cocks. Besides the gun, another mode of capturing the Woodcocks used till very lately to be, and perhaps still is, practised at Woodlands and some other places where practicable in Guernsey. Nets are set across open paths between the trees, generally Ilex, through which the Woodcocks take their flight when going out "roading," as it is called—that is, when on their evening excursion for food; into these nets the Woodcocks fly and become easy victims.

Professor Ansted includes the Woodcock in his list, but only marks it as occurring in Guernsey and Sark. There is one specimen in the Museum.

119. SOLITARY SNIPE. *Scolopax major*, Gmelin. French, "Grande becassine."—I have never been fortunate enough to shoot a Solitary

Snipe myself in the Channel Islands, neither have I seen one at any of the bird-stuffers; but that is not very likely, as the shooter of a Solitary Snipe only congratulates himself on having killed a fine big Snipe, and carries it off for dinner, but, from some of the descriptions I have had given me of these fine big Snipes, I have no doubt it has occasionally been a Solitary Snipe. Mr. MacCulloch also writes me word that the Solitary Snipe occasionally occurs.

It is included in Professor Ansted's list, and marked by him as occurring in Guernsey and Sark. There is no specimen at present in the Museum.

120. SNIPE. *Gallinago gallinaria*, Gmelin. French, "Bicassine ordinaire." — The Common Snipe is a regular and rather numerous autumnal visitor to all the Islands, remaining through the winter and departing again in the spring, some few remaining rather late into the summer. I am very sceptical myself about the Snipe breeding in the Channel Islands in the present day, although I was told one or two were seen about Mr. De Putron's pond late this summer, and were supposed to be breeding there; however, I could see nothing of them when there in June and July, although, as I have said before, Mr. De Putron kindly allowed me to search round his pond for either birds or eggs. Mr. MacCulloch, however, thinks they still breed in Guernsey, as he writes to me to say, "I believe that Snipes continue to breed here occasionally; I have heard of them, and put them up myself in summer." If they do, I should think the most likely places would be the wild gorse and heath-covered valleys leading down to the Gouffre and Petit Bo Bay, as there is plenty of water and soft feeding places in both; I have never seen one there, however, though I have several times walked both those valleys and the intervening land during the breeding-season, and I should think all these places were much too much overrun with picnic parties and excursionists to allow of Snipes breeding there now. Should the Snipe, however, still breed in the Island, it would be as well to give it a place in the Guernsey Bird Act, as it is much more worthy of protection during the breeding-season than many of the birds there mentioned. Sometimes in the autumn I have seen and

shot Snipe in the most unlikely places when scrambling along between huge granite boulders lying on a surface of hard granite rock, where it would be perfectly impossible for a Snipe to pick up a living; indeed with his sensitive bill I do not believe a Snipe, if he found anything eatable, could pick it off the hard ground. Probably the Snipes I have found in these unlikely places were not there by choice, but because driven from their more favourite places by the continual gunning going on in almost every field inland.

The Snipe is included in Professor Ansted's list, but only marked as occurring in Guernsey: it is difficult to say why this should be, when the Solitary Snipe and the Jack Snipe are marked as occurring in Guernsey and Sark, and all three are, at least, as common in Alderney as in the other two Islands. There is one specimen in the Museum.

121. JACK SNIPE. *Gallinago gallinula*, Linnaeus. French, "Bicassine Jourde."—The Jack Snipe is a regular autumnal visitant to all the Islands, but never so numerous as the Common Snipe. A few may always be seen, however, hung up in the market with the Common Snipes through the autumn and winter.

Professor Ansted includes it in his list, and marks it only as occurring in Guernsey and Sark. There is no specimen at present in the Museum.

122. KNOT. *Tringa canutus*, Brisson. French, "Becasseau canut," "Becasseau maubhche."—Common as the Knot is on the south and west coast of England during autumn and winter, it is by no means so common in the Channel Islands. I have never shot it there myself in any of my autumnal expeditions. Miss C.B. Carey records one, however, in the 'Zoologist' for 1871, as having been shot on September the 23rd of that year; and Mr. Harvie Brown mentions seeing a solitary Knot far out on the shore at Herm in January, 1869. These are the only occasions I am certain about, although it probably occurs sparingly every year, but I have never seen it even in the mar-

ket, and were it at all common a few certainly would have occasionally found their way there.

Professor Ansted includes it in his list, but only marks it as occurring in Guernsey. There is no specimen at present in the Museum.

123. CURLEW SANDPIPER. *Tringa subarquata*, G|ldenstaedt. French, "Becasseau cocorli." — The Curlew Sandpiper, or Pigmy Curlew as it is sometimes called, can only be considered a rare occasional visitant to the Channel Islands. I have never seen or shot one there myself, but Mr. Couch records one in the 'Zoologist' for 1874 as having been shot near the Richmond Barracks on the 5th of October of that year. Colonel L'Estrange told me also that some were seen in a small bay near Grand Rocque in the autumn of 1877. It may, however, have occurred at other times and been passed over or looked upon as only a Purre, from which bird, however, it may immediately be distinguished by its longer legs and taller form when on the ground, and by the white rump.

It is not included in Professor Ansted's list, and there is no specimen in the Museum.

124. PURRE or DUNLIN. *Tringa alpina*, Linnaeus. French, "Becasseau brunette," "Becasseau variable." — The Purre is resident in all the Islands throughout the year in considerable numbers, which however are immensely increased in the autumn by migratory arrivals, most of which remain throughout the winter, departing in the spring for their breeding stations. Though resident throughout the year, and assuming full breeding plumage, I am very doubtful as to the Purre breeding in the Islands; I have never been able to find eggs, nor, as a rule, have I found the bird anywhere but on its ordinary winter feeding-ground, amongst the mud and seaweed between high and low water mark. The most likely parts to find them breeding seem to be some of the high land and heather in Guernsey and the sandy common on the northern part of Herm, near which place I saw a few this summer (1878) in perfect breeding plumage,

and showing more signs of being paired than they generally do, and in parts of Alderney.

Professor Ansted has not mentioned it in his list. There are two specimens in the Museum, both in breeding plumage.

125. LITTLE STINT. *Tringa minuta*, Leishler. French, "Becasseau echasses," "Becasseau minute." — The Little Stint is only an occasional and never numerous autumnal visitant. I have seen one or two in the flesh at Mr. Couch's, killed towards the end of October, but I have never seen one alive or shot one myself.

It is included in Professor Ansted's list, and marked as occurring in Guernsey only. There is no specimen in the Museum.

126. SANDERLING. *Calidris arenaria*, Linnaeus. French, "Sanderling variable." — The Sanderling is a regular and rather early autumn visitant to all the Islands, as I have shot one as early as the end of August in Cobo Bay in Guernsey; this is about the time the Sanderling makes its first appearance on the opposite side of the Channel at Torbay. I have not met with it later on in October and November, but no doubt a few remain throughout the winter as they do in Torbay, where I have shot Sanderlings as late as the 27th of December; a few also probably visit the Islands on their return migration in the spring. The two in the Museum seem to bear out this, as one is nearly in winter plumage, and the other is assuming the red plumage of the breeding season, and could not have been killed before April or May.

The Sanderling is included in Professor Ansted's list, and marked by him as occurring in Guernsey and Sark.

127. GREY PHALAROPE. *Phalaropus fulicarius*, Linnaeus. French, "Phalarope gris," "Phalarope roussbtre," "Phalarope phatyrhinque." [19] — The Grey Phalarope is a tolerably regular and occasionally

numerous autumnal visitant to all the Islands, not, however, arriving before the end of October or beginning of November. At this time of year the greater numbers of birds are in the varied autumnal plumage so common in British-killed specimens, showing partial remains of the summer plumage; but one I have, killed in November, 1875, was in most complete winter plumage, there not being a single dark or margined feather on the bird. This perfect state of winter plumage is by no means common either in British or Channel Island specimens, so much so that I do not think I have seen one in such perfect winter plumage before.

The Grey Phalarope is included in Professor Ansted's list, but no letters marking its distribution through the Islands are added, perhaps because it was considered to be generally distributed through all of them. There is no specimen at present in the Museum.

128. HERON. *Ardea cinerea*, Linnaeus. French, "Heron cendri", "Heron huppi." — A good many Herons may be seen about the Islands at all times of the year; those that remain through the summer, though scattered over all the Islands, are probably all non-breeding birds. I have seen them fishing along the shore in Guernsey, Herm, Alderney, and the rocky islands north of Herm, but I have never seen or heard of an egg being found in either of the Islands, nor have I ever seen anything that bore the most remote resemblance to the nest of a Heron. Mr. MacCulloch, however, writes to me as follows: "The Heron is said to breed occasionally on the Amfrocques and others of those small islets north of Herm." Mr. Howard Saunders, Col. L'Estrange, and myself, however, visited all these islets this last breeding season (1878), and though we saw Herons about fishing in the shallow pools left by the tide, we could see nothing that would lead us to suppose that Herons ever bred there, in fact, though Herons have been known to breed on cliffs by the sea; the Amfroques and all the other little wild rocky islets are apparently the most unlikely places for Herons to breed on. In Guernsey itself, however, it is more likely that a few Herons formerly bred, and that there was once a small Heronry in the Vale. As Mr. MacCulloch writes to me, "There is a locality in the parish of St.

Samson, at the foot of Delancy Hill, in the vicinity of the marshes near the Ivy Castle, formerly thickly wooded with old elms, which bears the name of La Heronihre. It may have been a resort of Herons, but I am bound to say the name may have been derived from a family called 'Heron,' now extinct." It seems to me also possible that the family derived their name from being the proprietors of the only Heronry in Guernsey. In the place mentioned by Mr. MacCulloch there are still a great many elm trees quite big enough for Herons to build in, supposing they were allowed to do so, which would not be likely at the present time. The number of Herons in the Channel Islands seems to me to be considerably increased in the autumn, probably by wanderers from the Heronries on the south coast of Devon and Dorset; on the Dart and the Exe, and near Poole.

The Heron is included in Professor Ansted's list, but only marked as occurring in Guernsey. There is no specimen at present in the Museum.

129. PURPLE HERON. *Ardea purpurea*, Linnaeus. French, "Heron pourpre." — The Purple Heron is an occasional accidental wanderer to all the Islands. Mr. MacCulloch writes me word, "I have notes of that beautiful bird, the Purple Heron, being killed here (Guernsey) in May, 1845, and in 1849; also in Alderney on the 8th May, 1867." Curiously enough Mr. Rodd records the capture of one, a female, near the Lizard, in Cornwall, late in April of the same year. [20] When at Alderney this summer (1878) I was told that a Heron of some sort, but certainly not a Common Heron, had been shot in that Island about six weeks before my visit on the 27th of June. Accordingly I went the next morning to the bird-stuffer, Mr. Grieve, and there I found the bird and the person who shot it, who told me that it rose from some rather boggy ground at the back of the town — that he shot at it and wounded it, but it flew on towards the sea; and as it was getting rather late he did not find it till next morning, when he found it dead near the place he had marked it down the night before. It was in consequence of going to look up this bird that I found the Greenland Falcon before mentioned, which had been shot by the

same person. These are all the instances I have been able to collect of the occurrence of the Purple Heron in the Channel Islands.

It is, however, included in Professor Ansted's list, and marked as occurring in Guernsey, probably on the authority of one of the earlier specimens mentioned by Mr. MacCulloch. There is no specimen at present in the Museum.

130. SQUACCO HERON. *Ardeola cornuta*, Pallas. French, "Heron crabier." — I have in my collection a Guernsey-killed specimen of the Squacco Heron, which Mr. Couch informed me was shot in that island in the summer of 1867, and from inquiries I have made I have no doubt this information is correct. Mr. MacCulloch also writes to me to say, "A Squacco Heron was shot in the Vale Parish on the 14th of May, 1867, no doubt the one Couch sent to you." This was duly recorded by me in the 'Zoologist' for 1872, and is, I believe, the first recorded instance of its occurrence in the Channel Islands.

It is not mentioned in Professor Ansted's list, and there is no specimen in the Museum.

131. BITTERN. *Botaurus stellaris*, Linnaeus. French, "Heron grand butor," "Le grand butor." — Bitterns were probably at one time more common in Guernsey than they are at present, drainage and better cultivation having contributed to thin their numbers, as it has done in England; and Mr. MacCulloch tells me that in his youth they were by no means uncommon. Of late years, however, they have become much more uncommon, though, as he adds, specimens have been shot within the last three or four years. They seem now, however, to be confined to occasional autumnal and winter visitants. Mr. Couch says ('Zoologist' for 1871): — "On the 30th December, 1874, after a heavy fall of snow, I had a female Bittern brought to me to be stuffed, shot in the morning in the Marais; and on the 2nd of January following another was shot on the beach near the Vale Church. I had also part of some of the quill-feathers of a Bittern sent to me for identification by Mrs. Jago, which had been killed in

the Islands the last week in January, 1879." These are the most recent specimens I have been able to get any account of. The bird-stuffer in Alderney (Mr. Grieve) and his friend told me they had shot Bitterns in that island, but did not remember the date.

The Bittern is included in Professor Ansted's list, but only marked as occurring in Guernsey. There is no specimen in the Museum.

132. AMERICAN BITTERN. *Botaurus lentiginosus*, Montagu. French, "Heron lentigineux." [21] —This occasional straggler from the New World has once, in its wanderings, reached the Channel Islands, and was shot in Guernsey on the 27th October, 1870, and was duly recorded by me in the 'Zoologist' for 1871; it is now in my collection. This is the only occurrence of this bird in the Channel Islands yet recorded; but as the bird occasionally crosses to this side of the Atlantic—several specimens having occurred in the British Islands—it may possibly occur in Guernsey or some of the Channel Islands again. It may, therefore, be as well to point out the principal distinctions between this bird and the Common Bittern last mentioned. Between the adult birds there can be no mistake: the longer and looser feathers on the fore part of the neck, which are slightly streaked and freckled with dark brown, may be immediately distinguished from the much shorter and more regularly marked feathers on the neck of the adult American Bittern. This distinction, however, is not perfectly clear in young birds; but, at any age or in any state of plumage, the birds may be immediately distinguished by the primary quill-feathers, which in the American Bittern are a uniform dark chocolate-brown without any marks whatever, while in the Common Bittern they are much marked and streaked with pale yellowish brown; this may be always relied on at any age or in any plumage.

The American Bittern is not mentioned in Professor Ansted's list, no specimen having been found in the Channel Islands till after the publication of his list, and of course there is no specimen in the Museum.

133. LITTLE BITTERN. *Ardetta minuta*, Linnaeus. French, "Heron Blongios." [22] — I only know of one occurrence of the Little Bittern in the Channel Islands, and that was towards the end of November, 1876; and Mr. Couch writes to me as follows on the 3rd of December: "A very good Little Bittern was caught alive in the Vale Road; after being shot at and missed by two men, a young man in the road threw his pocket-handkerchief at it and brought it in to me alive." Mr. Couch also informed me, when he forwarded me the specimen, that it was a male by dissection. It is now in my collection, and is a young bird of the year. I am rather sorry that as Mr. Couch got it alive he did not forward it to me in that state, as, unless it had been wounded by the two shots, I have no doubt I should have been able to keep it alive and observe its habits and changes of plumage as it advanced towards maturity.

The Little Bittern is included in Professor Ansted's list, and marked as occurring in Guernsey. There is no specimen in the Museum.

134. SPOONBILL. *Platalea leucorodia*, Linnaeus. French, "Spatule blanche." — An occasional but by no means common visitor to the Channel Islands. I have been able to hear of but very few instances of its occurrence or capture of late years; Mr. Couch, however, writes me, in a letter dated November, 1873, that a Spoonbill was brought to him to stuff. In all probability this is the same bird recorded by Mr. Broughton in the 'Field' for October 25th, 1873, and in the 'Zoologist' for January, 1874. This is the only very recent specimen I have been able to trace; but Mr. Broughton in his note mentions the occurrence of one about twenty years before; and Mrs. Jago, who, when she was Miss Cumber, did a good deal of bird-stuffing in Guernsey, told me she had stuffed a Spoonbill for the Museum about twenty years ago. This is probably the other one mentioned by Mr. Broughton, and he may have seen it in the Museum; it is not there, however, now — either having become moth-eaten, and consequently thrown away, or lost when the Museum

changed its quarters across the market-place. Mr. MacCulloch does not seem to consider the Spoonbill such a very rare visitant to the Channel Islands, as he writes to me, "The Spoonbill is not near so rare a visitor as you seem to think; specimens were killed here in 1844, and in previous years, and again in 1849, and in October, 1873. [23] They are seldom solitary, but generally appear in small flocks. I forget whether it was in 1844 or 1849 that flocks were reported to have been seen in various parts of England, even as far west as Penzance. I think that in one of these years as many as a dozen were seen here in a flock." Mr. Rodd, in his 'List of the Birds of Cornwall,' does not mention either of these years as great years for Spoonbills, only saying, "Occasionally, and especially of late years, observed in various parts of the county; a flock of several was seen and captured at Gwithian; others have been obtained from the neighbourhood of Penzance, and also from Scilly." [24]

The Spoonbill is included in Professor Ansted's list, and marked as occurring in Guernsey. There is no specimen at present in the Museum, the one stuffed by Miss Cumber having, as above mentioned, disappeared.

135. WHITE-FRONTED GOOSE. *Anser albifrons*, Scopoli. French, "Oie rieuse, ou ` front blanc."—None of the Grey Geese seem common in Guernsey; neither the Greylag, the Bean, nor the Pink-footed Goose have, as far as I am aware, been obtained about the Islands, nor have I ever seen any either alive or in the market, where they would be almost sure to be brought had they been shot by any of the fishermen or gunners about the Islands. There is one specimen, however, of the White-fronted Goose in the Museum, which I have reason to believe was killed in or near Guernsey; and this is the only specimen of this Goose which, as far as I am aware, has been taken in the Islands.

The White-fronted Goose is included in Professor Ansted's list, and marked as occurring in Guernsey. The Greylag and the Bean Goose are also included in the list, the Greylag marked as occurring in Guernsey and Sark, and the Bean as only in Guernsey; but no information beyond the letter marking the locality is given as to either; and the only specimen in the Museum is the White-fronted Goose above mentioned, neither of the others being represented there now, nor do I remember ever having seen a specimen of either there.

136. BRENT GOOSE. *Bernicla brenta*, Brisson. French, "Oie cravant," "Bernache cravant."—The Brent Goose is a regular winter visitor to all the Islands, varying, however, in numbers in different years: sometimes it is very numerous, and affords good sport during the winter to the fishermen, who generally take a gun in the boat with them as soon as the close season is over, sometimes before. The flocks generally consist mostly of young birds of the year; the fully adult birds, however, though fewer in number, are in sufficient numbers to make a very fair show.

Professor Ansted includes it in his list, but only marks it as occurring in Guernsey and Sark; it is, however, quite as common about Herm and Alderney. There is no specimen at present in the Museum.

137. MUTE SWAN. *Cygnus olor*, Linnaeus. French, "Cygne tuberculi."—I do not believe this bird has ever visited the Channel Islands in a thoroughly wild state, though it is pretty widely spread over Europe; its range, however, being generally more to the east than the Channel Islands. Mr. Couch, however, at page 4939 of the 'Zoologist' for 1874, records the occurrence of two Mute Swans on the 7th of September at the Braye Pond, where they were shot. He also says that "five others passed over the Island the same day; they were flying low, and, judging from their colour, were young birds." As no one in the Islands keeps Swans, these were most probably a family party that had strayed away from the Swannery at Abbotsbury, on the opposite coast of Dorset, where some three hundred and fifty pairs still breed annually. I have myself seen as many six hundred and thirty birds there, the hens sitting and the old males each resting quietly by the nest, keeping guard over the fe-

male and the eggs. The distance from the Abbotsbury Swannery, which is at the extreme end of the Chesil Beach, in Dorsetshire, to Guernsey is nothing great for Swans to wander; and they often, both old and young (after the young are able to fly), wander away from their home as far as Exmouth on one side and Weymouth Bay or the Needles on the other; and an expedition to Guernsey would be little more than to one of these places, and by September the young, which are generally hatched tolerably early in June (I have seen a brood out with their parents on the water as early as the 27th of May), would be perfectly able to wander, either by themselves or with their parents, as far as the Channel Islands, and, as at this time they rove about outside the Chesil Beach a good deal, going sometimes a long way out to sea, there is no reason they should not do so. It seems a great pity that these fine birds should be shot when they wander across channel to Guernsey, especially when it must be apparent to every one that they are really private property. If the present long close season is to be continued, the Mute Swan might well be added to the somewhat unreasonable list of birds in the Guernsey Sea-birds Act; at all events, Swans would be better worth preserving than Plongeons or Cormorants.

138. HOOPER. *Cygnus musicus*, Bechstein. French, "Cygne sauvage."—The Wild Swan or Hooper [25] is an occasional visitor to the Channel Islands in hard winters, sometimes probably in considerable numbers, as Mrs. Jago (late Miss Cumber) told me she had had several to stuff in a very hard winter about thirty years ago; some of these were young birds, as she told me some were not so white as others. Mr. MacCulloch also says that the Hooper visits the Channel Islands in severe winters; and the capture of one is recorded by a correspondent of the 'Guernsey Mail and Telegraph' for 4th January, 1879, as having been shot in that Island a few days before; it is said to have been a young bird, grey in colour. The writer of the notice, while distinguishing this bird from the Mute Swan, does not, however, make it so clear whether it was really the present species or Bewick's Swan; from the measurement of the full length (5 ft. 3 in.) given, however, it would appear that it was the present species, as that would be full length for it, while Bewick's Swan would be

about one-third less; some description of the bill, however, would have been more satisfactory. It would certainly have been interesting to have had some more particulars about this Swan, as this last severe winter (1878 and 1879) has been very productive of Swans in the south-west of England, the greater number of those occurring in this county of Somerset, however, curiously enough, having been Bewick's Swan, which is generally considered the rarer species. Though Swans have been so exceptionally numerous in various parts of England this winter, the above-mentioned is the only occurrence I have heard of in the Channel Islands.

The Hooper is included in Professor Ansted's list, but marked as only occurring in Guernsey. There are two specimens in the Museum, one adult and one young bird.

139. BEWICK'S SWAN. *Cygnus minor*, Keys and Blasius. French, "Cygne de Bewick." [26] —I have very little authority for including Bewick's Swan in my list of Guernsey birds; Mr. MacCulloch, however, writes me word, "The Common Hooper has visited us in severe winters, and is certainly not the *only* species of *wild* Swan that has been shot here." In all probability the other must have been Bewick's Swan, which no doubt has occasionally occurred, perhaps more frequently than is supposed, though not so frequently as the Hooper. Probably the difference between the two is not sufficiently known; it may, therefore, be as well to point out the distinctions. Bewick's Swan is much smaller than the Hooper, but the great outward distinction is, that in the Hooper the yellow at the base of the bill extends to and includes the nostrils, whereas in Bewick's Swan the yellow occupies a very small portion of the base of the bill, not extending so far as the nostrils: this is always sufficient to distinguish the two, and is almost the only exterior distinction, but on dissection the anatomical structure, especially of the trachea, shows material difference between the two.

Professor Ansted includes Bewick's Swan in his list, and marks it as occurring in Guernsey. There is, however, no specimen at present in the Museum.

140. WILD DUCK. *Anas boschas*, Linnaeus. French, "Canard sauvage."—-The Wild Duck is an occasional autumn and winter visitant. I have never shot one myself in the Islands, but I have several times seen Guernsey-killed ones in the market. Though a visitant to all the Islands, I do not believe the Wild Duck breeds, at all events at present, in any of them; Mr. MacCulloch, however, writes me word "The Wild Duck formerly bred here;" and Mr. Gallienne, in his 'Notes' to Professor Ansted's list, says—"The Wild Duck formerly bred in Guernsey rather abundantly, but it seldom does so now. Last year a nest was found on one of the rocks near Herm." This would be about 1861. The rocks to the northward of Herm do not seem to me a likely place for the Wild Duck to breed; however, there are one or two places where they might possibly do so. A much more likely place would be in some of the reed beds in the Grande Mare, or even amongst the heather and gorse above the high cliffs on the south and east side of the Island,—a sort of place they are fond of selecting in this county, Somerset, where they frequently nest amongst the heather high up in the hills, and quite away from any water.

The Wild Duck is included in Professor Ansted's list, and marked as occurring in Guernsey and Sark. There is no specimen at present in the Museum.

141. PINTAIL. *Dafila acuta*, Linnaeus. French, "Pilet," "Canard pilet." The Pintail is an occasional autumn and Winter visitant, but never very common. I have one specimen, a female, killed in Guernsey in November, 1871, and this Mr. Couch told me was the only one he had had through his hands whilst in Guernsey; and Captain Hubbach writes me word that he shot one in Alderney in January, 1863. I have never seen it in the Guernsey market, like the Wild Duck and Teal.

Professor Ansted includes it in his list, but only marks it as occurring in Guernsey. There is one specimen, a male in full plumage, in the Museum.

142. TEAL. *Querquedula crecca*, Linnaeus. French, "Sarcelle d'hiver."—Like the Wild Duck, the Teal is a regular but never numerous visitant to all the Islands. A few make their appearance in the Guernsey market in October and November, and occasionally through the winter; but Teal do not, as a rule, add much to the Guernsey sportsman's bag. In November, 1871, a friend of mine told me that, after a long day's shooting from daylight till dark, he succeeded in bagging one Teal and one Woodcock. I was rather glad I was not with him on this occasion, but chose the wild shooting on the shore, where I got one or two Golden Plovers, and Turnstone and Ring Dotterel enough for a pie—and, by-the-bye, a very good pie they made.

Professor Ansted includes the Teal in his list, and marks it as occurring in Guernsey and Sark. There is no specimen in the Museum at present.

143. EIDER DUCK. *Somateria mollissima,* Linnaeus. French, "Canard eider," "Morillon eider."—The Eider Duck occasionally straggles to the Channel Islands in the autumn, but very seldom, and the majority of those that do occur are in immature plumage. I have one immature bird, killed in Guernsey in the winter of 1876; and that is the only Channel Island specimen that has come under my notice, and I think almost the only one Mr. Couch had had through his hands.

The Eider Duck is included in Professor Ansted's list, and marked as occurring in Guernsey. The King Eider is also included in the list, but no letter marking the distribution through the Islands is given, and no information beyond the mere name, so I should think in all probability this must have been a mistake, especially as I can find no other evidence whatever of its occurrence. There is no specimen of either bird in the Museum.

144. COMMON SCOTER. *Oidemia nigra*, Linnaeus. French, "Macreuse," "Canard macreuse."—The Scoter is a common autumn and winter visitor to all the Islands, generally making its appearance in considerable flocks; sometimes, however, the flocks get broken up, and single birds may then be seen scattered about in the more sheltered bays. Some apparently remain till tolerably late in the spring as Mr. MacCulloch wrote me word that a pair of Scoters were killed in the last week in April, 1878, off the Esplanade; he continues, "I had only a cursory glance of them as I was passing through the market in a hurry, and I am not sure they were not Velvet Scoters. The male had a great deal of bright yellow about the nostrils." Mr. MacCulloch, however, told me afterwards, when I asked him more about them, and especially whether he had seen any white about the wing, that he had not seen any white whatever about them, so I have but little doubt that they were Common Scoters, and he could hardly have failed to be struck by the conspicuous white bar on the wing, by which the Velvet Scoter, both male and female, may immediately be distinguished from the Common Scoter. As on the South Coast of Devon or Dorset, a few scattered Scoters—non-breeding birds, of course—remain throughout the summer. I have one, a male, killed off Guernsey on July 19th: this bird is in that peculiar state of plumage which all the males of the *Anatidae* put on from about July to October, and in which many of them look so like the females.

The Common Scoter is included in Professor Ansted's list, and marked only as occurring in Guernsey. The Velvet Scoter is also included in Professor Ansted's list, and marked as occurring in Guernsey; but there seems to be no other evidence of its having occurred in the Islands; and a mistake may easily have been made, however, as the Velvet Scoter occurs tolerably frequently on the south coast of Devon, though never in such numbers as the Common Scoter; it may, of course, occur in the Channel Islands occasionally. There is no specimen of either bird in the Museum.

145. GOOSANDER. *Mergus merganser*, Linnaeus. French, "Grand Harle."—The Goosander is a regular and tolerably numerous visi-

tant to all the Islands, arriving in the autumn and remaining throughout the winter. The heavy-breaking seas of the Channel Islands do not appear to disturb the composure of these birds in the least, for once, on my voyage home on the 16th November, 1871, I saw a small flock of Goosanders off Herm, close to the steamer; they were swimming perfectly unconcerned in a heavy-breaking sea, which made the steamer very lively, dipping first one and then the other paddle-box into the water; as we got close up to them they rose, but only flew a short distance and pitched again in the white water. They seem to me to keep the sea better than the Red-breasted Merganser—at least, I have not seen them seek shelter so much in the different bays.

The Goosander is included in Professor Ansted's list, but only marked as occurring in Guernsey. There is no specimen in the Museum at present, though I think there used to be one, but I suppose it has got moth-eaten and been thrown away.

146. RED-BREASTED MERGANSER. *Mergus serrator*, Linnaeus. French, "Harle Huppi."—Like the Goosander, the Red-breasted Merganser is a regular and by no means uncommon autumn and winter visitant to the Channel Islands. It seems to me, as I said before, that these birds seek the more sheltered bays during wild squally weather more than the Goosanders do; not but what they can keep the sea well even in bad weather, but I have never seen or shot the Goosander close to the shore seeking smooth water, as I have done the Red-breasted Merganser. The greater number of Red-breasted Mergansers killed in the Channel Islands which I have seen have been either females or males that had not assumed the full adult plumage—in fact, in that state of plumage in which they are the "Dun Diver" of Bewick; full-plumaged adult males do, however, occur as well as females and young males, or males in a state of change.

Professor Ansted includes the Red-breasted Merganser in his list, but only marks it as occurring in Guernsey. There are two specimens in the Museum—a male in full plumage and a female or young male.

147. SMEW. *Mergus albellus*, Linnaeus. French, "Harle piette," "Harle itoili," "Petit harle huppi."—The Smew can only be considered an occasional accidental autumnal visitant, and the few that do occur are generally either females, young males, or males still in a state of change. I do not know of any instance in which a full-plumaged male has occurred in the Channel Islands.

It is mentioned in Professor Ansted's list, and marked as occurring in Guernsey only. There are two specimens in the Museum, both females or immature males, or, at all events, males which have not begun to assume their proper plumage after the summer change.

148. LITTLE GREBE. *Podiceps minor*, Gmelin. French, "Grhbe castagneux."—The Little Grebe, or Dabchick, occurs occasionally in the Islands, mostly as an autumnal or winter visitant. I have occasionally seen freshly-killed ones hanging up in the market in November; I have, however, never seen it alive or shot it in the Islands. Mr. Couch, writing to me in December, 1876, told me that Mr. De Putron had told him that Little Grebes had bred in his pond in the Vale the summer before, and Mr. De Putron afterwards confirmed this; they can only breed there occasionally, however, as there were certainly none breeding there in 1878, when I was there.

The Little Grebe is included in Professor Ansted's list, and marked by him as occurring in Guernsey only. There are two specimens in the Museum and some eggs, which were said to be Guernsey, and probably were so, perhaps from the Vale Pond.

149. EARED GREBE. *Podiceps nigricollis*, Sundeval. French, "Grhbe oreillard."—The Eared Grebe is an occasional autumnal visitant to the Islands, remaining on till the winter; it is never very numerous; in some years, however, it appears to visit the Islands in greater numbers than in others, as Mr. Couch mentions, at p. 4380 of the

'Zoologist' for 1875, that, amongst other grebes, four Eared Grebes were brought to him between the 4th and 13th of January. I do not know, however, that it ever occurs at any time of year except the winter and autumn; and I have never seen a Channel Island specimen in breeding plumage, or even in a state of change.

The Eared Grebe is included in Professor Ansted's list, but only marked as occurring in Guernsey. There is now no specimen in the Museum.

150. SCALAVONIAN GREBE. *Podiceps auritus,* Linnaeus. French, "Grhbe cornu ou Esclavon." — The Sclavonian Grebe is a regular and rather numerous autumn and winter visitor to all the Islands. In rough weather it may be seen fishing about the harbour at Guernsey when it can find any protection from the rough seas that so often rage all round the Island, and which drive it to seek shelter either about the harbour or some of the more protected bays. I do not know that it has ever bred in the Islands, but there was a very fine specimen in full breeding-plumage at the late Mr. Mellish's, which I often saw there; and, on subsequent inquiry from his son, Mr. William Mellish, he wrote in 1878 to me to say, "The Sclavonian Grebe was killed by my brother Alfred at Arnold's Pond, just the other side of the Vale Church to the one on which you were." This Arnold's Pond is the one I have so often mentioned before as Mr. De Putron's. I have not been able to ascertain the exact date at which this bird was killed, but it must have been some time in the spring, as it was in full breeding-plumage. There is also one in full breeding-plumage in the Museum, so it must occasionally stay on some time into the spring. The young birds and adults in winter plumage, when it is the Dusky Grebe of Bewick, are very much like the Eared Grebe in the same state of plumage; but they may always be distinguished, the Sclavonian Grebe always being rather the larger and having the bill straighter, and making a more regular cone than that of the Eared Grebe, which is slightly turned up. In the full breeding-plumage there can be no possibility of confounding the two species.

The Sclavonian Grebe is included in Professor Ansted's list, but only marked as occurring in Guernsey. There are two specimens in the Museum, one in full breeding-plumage and one in winter plumage.

151. RED-NECKED GREBE. *Podiceps griseigena,* Boddaert. French, "Grhbe jou-gris."—I have never seen a Channel Island specimen of the Red-necked Grebe in full breeding-plumage as I have the Sclavonian, but it is a tolerably regular autumn and winter visitant, and in some years appears to be the more numerous of the two. Certainly in November, 1875, this was the case, and the Red-necked Grebe was commoner than either the Great-crested or the Sclavonian Grebe, especially about the Guernsey coast between St. Peter's Port and St. Samson's, where I saw several; and a good many were also brought into Mr. Couch's about the same time more than usual. One which I obtained had slight traces of the red about the throat remaining, otherwise this one was like the others which I saw in complete winter plumage.

The Red-necked Grebe is included in Professor Ansted's list, but only marked as occurring in Guernsey. There is one specimen in the Museum.

152. GREAT-CRESTED GREBE. *Podiceps cristatus,* Linnaeus. French. "Grhbe huppi."—The Great-crested Grebe is a regular autumn and winter visitant to the Channel Islands, but not, I think, in quite such numbers as at Teignmouth and Exmouth and along the south coast of Devon. I have not shot this bird in the Channel Islands myself, nor have I seen it alive: but I have seen several Guernsey-killed specimens. These were all young birds or adults in winter plumage; and I have one, a young bird of the year, killed in the Guernsey harbour late in November, 1876.

It is included in Professor Ansted's list, but only marked as occurring in Guernsey. There is one specimen, a young bird of the year, in the Museum.

153. GREAT NORTHERN DIVER. *Colymbus glacialis*, Linnaeus. French, "Plongeon imbrim."—The Great Northern Diver is a common autumn and winter visitant to all the Islands, arriving early in November, perhaps even about the last week in October. The earliest date at which I have seen it myself was on the 9th November. A considerable majority of these autumnal visitants are young birds of the year, the rest being adults in winter plumage; but, as is the case on the south coast of Devon, a few occasionally remain so late on in the spring as to have fully attained the breeding-plumage. There is one Guernsey-killed specimen in perfect, or nearly perfect, breeding-plumage in the Museum, which I think was killed some time in May by Mr. Peter Le Newry, a well-known fisherman and gunner living in Guernsey, who procured a good many specimens for that establishment, but, unluckily, no note as to date or locality has been preserved; he told me he had killed this bird late in the spring, but could not when I saw him remember the exact date. It must not be supposed that because this bird occasionally remains in the Islands late into the spring, and assumes its full breeding-plumage before leaving, that it ever remains to breed or avails itself of the protection so kindly afforded to it and its congeners, as well as their eggs, by the Guernsey Bird Act.

The Great Northern Diver is included in Professor Ansted's list, but only marked as occurring in Guernsey. There are four specimens in the Museum in full breeding plumage and change.

154. BLACK-THROATED DIVER. *Colymbus arcticus*, Linnaeus. French, "Plongeon ` gorge noir."—The Black-throated Diver is a much less common visitor to the Islands than either the Great Northern or Red-throated Diver; it does, however, occasionally occur in the autumn and winter; all the specimens that have been obtained are either immature or in winter plumage, and I do not know of a single instance in which it has been procured in full plumage as the Great Northern has. In the 'Zoologist' for 1875 Mr. Couch records the occurrence of a Black-throated Diver on the 19th of January of that year, and of another on the 30th of the same month; these are the most recent occurrences of which I am aware.

No doubt the young Black-throated Diver may be occasionally mistaken for and passed over as the young Northern Diver; but it may always be known by its much smaller size, being intermediate between that bird and the Red-throated Diver, from which, however, it may always be distinguished by wanting the white spots on the back and wing-coverts which are always present in the winter plumage of the adult Red-throated Diver, and the oval marks on the margins of the feathers of the same parts in the young birds of the year.

The Black-throated Diver is included in Professor Ansted's list, and marked as only occurring in Guernsey. There is one specimen, an immature bird, in the Museum.

155. RED-THROATED DIVER. *Colymbus septentrionalis*, Linnaeus. French, "Plongeon ` gorge rouge," "Plongeon cat-marin."—The Red-throated Diver is a regular autumn and winter visitant to the Islands, and rather the most common of the three Divers. As with the Northern Diver, it occasionally remains until it has nearly assumed its full breeding-plumage, but it does not occur so frequently in that plumage as it does on the south coast of Devon and Dorset; indeed I have never found either this bird or the Great Northern Diver so common in the Channel Islands as they are about Exmouth and Teignmouth, even in the ordinary winter plumage; probably the mouths of rivers were more attractive to them as producing more food than the wild open seas of the Channel Islands. Owing to its various changes of plumage, from age or time of year, the Red-throated Diver has been made to do duty as more than one species, and is the Speckled Diver of Pennant, Montagu and Bewick.

It is mentioned in Professor Ansted's list, but marked as only occurring in Guernsey. There is no specimen at present in the Museum.

156. GUILLEMOT. *Alca troile*, Linnaeus. French, "Guillemot ` capuchon," "Guillemot troile."—The Guillemot is very common

about the Channel Islands in Autumn and winter, but is seldom seen during the summer season except near its breeding stations, which, as far as my district is concerned, are very few. It does not breed in Guernsey, Sark, or Herm, or even on the rocky islands to the north of Herm. In Alderney, I am told, it has one small station on the mainland on the side nearest the French coast. I was told of this by the person who shot the Greenland Falcon, and by one or two of the fishermen on my last visit to that Island. I had not time then to visit the place, and on former visits I must quite have overlooked it. Captain Hubbach, however, kindly promised that he would visit the spot, and soon after I left, about the middle of June, 1878, he did so, and his account to me was as follows: — "I have been twice along the cliffs with my glass, but have not seen either a Guillemot or Razorbill. An old boatman here tells me that he took their eggs off the rocks at the French side of Alderney last year (1877), and that they bred there every year. He describes the eggs as 'the same blue and green and white ones with black spots that are on the Ortack Rock.'" This very much confirms what Mr. Gallienne says, in his notes to Professor Ansted's list—"The Razorbill and Guillemot breed on the Ortack Rock and on the cliffs at Alderney." This Ortack Rock is to the west of Alderney, between Burhou and the Caskets, and a considerable number of Guillemots and Razorbills breed there, but it is not to be compared as a breeding station for these birds with those at Lundy Island and South Wales. During the summer a few Guillemots, probably non-breeding birds, may be seen at sea round Guernsey, and one or two stragglers may generally be seen when crossing from Guernsey to Sark or Herm. I have never seen the variety called the Ringed Guillemot, *Alca lacrymans*, in the Channel Islands, but, as it may occasionally occur, it is as well to mention it, although it is now rightly considered only a variety of the Common Guillemot, from which it differs only in summer plumage, when it has a white ring round the eye, and a white streak passing backwards from the eye down the side of the neck: this distinction is not apparent in the winter plumage, nor is there any distinction between the eggs.

The Guillemot is included in Professor Ansted's list, but is only marked as occurring in Guernsey and Sark. There are two specimens in summer plumage in the Museum, and one in winter plumage.

157. LITTLE AUK. *Mergulus alle*, Linnaeus. French, "Guillemot nain." — The Little Auk can only be considered a rare occasional wanderer to the Channel Islands, generally driven before the heavy autumnal and winter gales. I only know of the occurrence of two specimens: one of these was recorded by Mr. Couch in the 'Zoologist' for 1875, as having been killed on the 30th January in that year; and I had a letter from Mr. Couch, dated the 20th December, 1872, in which he informed me that a Little Auk had been taken alive in Guernsey on the 17th of that month: this one had probably, as is often the case, been driven ashore during a gale, and, being too exhausted to rise, had been taken by hand.

The Little Auk is included in Professor Ansted's list, and marked as occurring in Guernsey and Sark. There is no specimen at present in the Museum.

158. PUFFIN. *Fratercula arctica*, Linnaeus. French, "Macareux." — The Puffin, or Barbelote [27] as it is called by the Guernsey sailors and in the Guernsey Bird Act, is a regular and numerous summer visitant to the Islands, breeding in considerable numbers in many places. None breed, however, in Guernsey itself, or in any of the little rocky islands immediately surrounding it. Some breed on Sark and the islands about it, and a few also on Herm; but their great breeding quarters about these parts are from the Amfrocques to the north end of Herm. On every one of the little rocky islands between these places, and including the Amfrocques, considerable numbers of Puffins breed, either in holes in the soft soil which has accumulated on some of these islands, or amongst the loose rocks and stones; these latter, however, are the safest places for the Puffin, as, in spite of the Guernsey Bird Act, which protects the eggs as well as the birds, the Guernsey fishermen are fond of visiting these islands whenever they can for the purpose of what they call "Barbeloting;"

and they soon lift up the loose earth with their hands and get at the eggs; but the Puffins, who have laid in holes in the rocks and amongst loose stones, are much better off, as a good big stone of two or three tons is not so easily moved. I visited all these little islands in the summer of 1878 with Mr. Howard Saunders, and we found all the Puffins who had had eggs in holes in the earth had been robbed almost without an exception; the others, however, were pretty safe. Besides these islands the Puffins breed in Alderney itself, and on Burhou, where, however, their eggs are robbed nearly as much as in the islands north of Herm, especially the eggs of those who choose holes in the soft earth. The Puffins do not seem to be very regular in their time of nesting; at least, when I was at Burhou on the 14th of June, 1876, I found quite fresh eggs, eggs just ready to hatch, young birds in the down, and young birds just beginning to get a few feathers and almost able to take to the water; it was fun to see one of these when he had been unearthed waddle off to the nearest hole as fast as his legs could carry him – generally, however, coming down every second or third step. The reason for the irregularity in hatching was probably owing to the first brood having been lost, the eggs probably having been robbed. During the breeding season the Puffins keep very close to their breeding-stations, and do not apparently wander more than a few hundred yards from them even in search of food; so that, unless you actually visit the islands on which they breed, you can form no idea of the number of Puffins actually breeding in the Channel Islands. The number of Puffins, however, at Burhou seem to me to have considerably diminished of late years, for in the summer of 1866, when going through the Swinge, we passed a great flock of these birds; "in fact, for more than a mile both air and water were swarming with them." [28] This certainly was not the case in either 1876 or 1878, though there were still a great many Puffins there; probably the continued egg-stealing has had some effect in reducing their numbers. After the breeding-season the Puffins seem to leave the Channel Islands for the winter, as they do at Lundy Island and in the British Channel; they may return occasionally, as they do in the Bristol Channel, for a short time in foggy weather; but I have never seen a Puffin in any of my passages in October and November, or in any boating expedition at that time of year, and I have never heard any of the boatmen talk about Barbelotes being seen about in the winter. An unsigned paper, however, in the 'Star' for April 27th, 1878, mentions Puffins amongst other winter birds; but I very much doubt their making their appearance in the winter except as accidental visitants; there is one specimen, however, in the Museum, which, judging by the bill, must have been killed in the winter, or, at all events, to quote Dr. Bureau, "aprhs la saison des amours." Dr. Bureau, in a very interesting paper [29] on this curious change, or rather moult, which takes place in the bill of the Puffin, and which has been translated into the 'Zoologist' for 1878, where a plate showing the changes is given, says that Puffins are cast ashore on the coast of Brittany during the winter, for he says they leave the coast, as I believe they do that of the Channel Islands, and the only indication of their continuing there is that dead birds are rolled on the shore after severe gales in the autumn and winter; and "these birds are clad in a plumage different to that worn by those we get in the breeding-season. In the orbital region, for instance, they have a spot, more or less large, of a dusky brown; they have not the red eyelids, nor the horny plates above and below the eye, nor have they the puckered yellow skin at the base of the bill, and, what is still more remarkable, the bill is differently formed; it is neither of the same size, shape, nor colour, and the pieces of which it is composed are not even the same. It is small sliced off (trongui) in front, especially at the lower mandible, wanting the pleat (ourlet) at the base, and flattened laterally on a level with the nostrils, where a solid horny skin of a bright lead-colour is replaced by a short membrane." The whole paper by Dr. Bureau on this subject is most interesting, but is much too long for me to insert here; the nature, however, of the change which takes place must be so interesting to many of my readers who are familiar with the Puffin in its breeding plumage, and who, in spite of the Bird Act, perhaps occasionally enjoy a day's "Barbeloting," that I could not help quoting as much of the paper as would be sufficient to point out the general nature of the change.

The Puffin is included in Professor Ansted's list, but marked as occurring only in Guernsey and Sark. There are two specimens in the Museum; one in the ordinary summer plumage, and one apparently in the winter plumage above described; but it is difficult to be quite certain on the subject, as it has been smeared over with bird-stuffer's paint, probably with the view of making it as like the ordinary summer plumage as possible.

159. RAZORBILL. *Alca torda*, Linnaeus. French, "Pingouin macroptere." — The Razorbill is not by any means numerous in the Channel Islands, but a few breed about Ortack, and, as has been said before, in Alderney, but nowhere else; and they are by no means so numerous as the Guillemot. It is resident throughout the year, though perhaps more common in the autumn than at any other time. Mr. Harvey Brown, [30] however, mentions seeing a small flock swim by with the tide, at the north-end of Herm, in January. Mr. MacCulloch writes me word he has a note of a Razorbill Auk shot in Guernsey on the 14th February, 1847; this, of course, is only a young Razorbill of the previous year, which had not at that time fully developed its bill.

The Razorbill is included in Professor Ansted's list, but only marked as occurring in Guernsey. There are two Razorbills in the Museum, one in summer and one in winter plumage.

160. CORMORANT. *Phalacrocorax carbo*, Linnaeus. French, "Grand cormoran." — The Cormorant is by no means common in the Islands; I have never seen it about Guernsey, though I have seen one or two near Herm; I do not know that it breeds anywhere in the Islands, except at Burhou, and there only one or two pairs breed. I was shown the nesting-place just at the opening of a small sort of cavern; there was, however, only the remains of one egg that had been hatched, and probably the young gone off with its parents. I, however, received an adult bird and a young bird of the year, shot in the harbour at Alderney in August of that year, and those are the only Channel Island specimens of the Cormorant that I have seen.

Professor Ansted includes the Cormorant in his list, and marks it as occurring only in Guernsey and Sark. There is no specimen at present in the Museum.

161. SHAG. *Phalacrocorax graculus*, Linnaeus. French, "Cormoran largup."—The Shag almost entirely takes the place, as well as usurps the name, of its big brother, as in the Islands it is invariably called the Cormorant. The local Guernsey-French name "Cormoran" is applicable probably to either the Shag or the Cormorant. The Shag is the most numerous of the sea birds which frequent the Islands, the Herring Gull not even excepted, every nook and corner of the high cliffs in all the Islands being occupied by scores of Shags during the breeding-season. They take care, however, to place their nests in tolerably inaccessible places that cannot well be reached without a rope. The principal breeding-places are—in Guernsey, about the Gull Cliffs, and from there to Petit Bo, and a few, but not so many, on the rocks between there and Fermain, wherever they can find a place; none breed on the north or west side of the Island; in Jethou and Herm, and on the rock called La Fauconnihre, a few also breed, but not so many as in Guernsey, and we did not find any breeding on the Amfrocques or the other rocks to the north of Herm. On Sark they breed in great numbers, mostly on the west side nearest to Guernsey, and on the Isle de Marchant or Brechou, especially on the grand cliffs on both sides the narrow passage which divides that Island from the mainland of Sark, and from there to the Coupie, and from there round Little Sark to the Creux Harbour on the south-east. On the east side, that towards the French coast, there are few or none breeding, the cliffs not being so well suited to them; a great number breed also on Alderney, on the high cliffs on the south and east, but none on Burhou. The Shags appear to breed rather earlier than the Herring Gulls; when I was in the Islands in June, 1876, almost all the Shags had hatched, and the young were standing by their parents on the rocks close to their nests. When I visited some of the breeding-places of the Shags on the 27th of May, 1878, neither Gulls nor Shags had hatched, but when I went to the Gull Cliff on the 20th of June I found nearly all the Shags had hatched, though none or very few of the Herring Gulls had done so; some of the young Shags had left the nests and were about on the water; others were nearly ready to leave, and several were little things quite in the down. Though it is generally

easy to look down upon the Shags on their nests, and to get a good view at a short distance of the eggs and the young, it is, as a rule, by no means easy to get at them without a rope; in a few places, however, their nests are more accessible, and a hard climb on the rocks, perhaps with a burning sun making them almost too hot to hold, will bring you within reach of a Shag's nest; but I would not advise any one who tries it to put on his "go-to-meeting clothes," as the deposit of guano on the rocks will spoil anything; and only let him smell his hands after his exploit—they do smell so nice! One of the parents generally stands by the young after they are hatched, I suppose to prevent them from wandering about and falling off the rocks, as the positions of some of them seem very critical, there being only just room for the family to stand; the other parent is generally away fishing, only returning at intervals to feed his family and dry his feathers before making a fresh start; sometimes one parent takes a turn to stay by the young, and sometimes the other. The usual number of young appeared to be three, sometimes only one or two; but in these cases it is probable that a young one or two may have waddled off the rock, or got into a crevice from which the parents could not extricate it, accidents which I should think frequently happen; or an egg or two may have been blown from the nest, or egg or young fallen a victim to some marauding Herring Gull during the absence of the parents. The Shag assumes its full breeding-plumage and crest very early; I have one in perfect breeding-plumage, killed in February; and Miss C.B. Carey mentions in the 'Zoologist' having seen one in Mr. Couch's shop with its full crest in January. I do not quite know at what time the young bird assumes adult plumage, but I have one just changing from the brown plumage of the young to adult plumage. Many of the green feathers of the adult are making their appearance amongst the brown ones; this one I shot on the 26th June, 1866, near the harbour Goslin, at Sark, near a large breeding-station of Shags and Herring Gulls: if it is, as I suppose, a young bird of the year, it would show a very early change to adult plumage, but of course it might have been a young bird of the previous year; but, as a rule, young birds of the previous year are not allowed about the breeding-stations, any more than they are by the Herring Gulls.

The Shag is included in Professor Ansted's list, but curiously enough only marked as occurring in Guernsey. There are two adult specimens and one young bird and one young in down in the Museum.

162. GANNET. *Sula bassana*, Linnaeus. French, "Fou de bassan." — The Gannet, or Solan Goose, as it is sometimes called, is a regular autumn and winter visitant to all the Islands, but never so numerous, I think, as on the south coast of Devon; birds, however, in all states of plumage, young birds as well as adults, and in the various intermediate or spotted states of plumage, make their appearance. It stays on through the winter, but never remains to breed as it does regularly at Lundy Island. I have seen both adults and young birds fishing round Guernsey, and Mrs. Jago (late Miss Cumber) told me she had had several through her hands when she was the bird-stuffer there; she also wrote to me on the 16th March, 1879, to say a fully adult Gannet had been shot in Fermain Bay on the 15th; and Mr. Grieve, the carpenter and bird-stuffer at Alderney, had the legs and wings of an adult bird, shot by him near that Island, nailed up behind the door of his shop. I do not think, however, that the strong tides, rough seas, and sunken rocks of the Channel Islands suit the fishing operations of the Gannet as well as the smoother seas of the south coast of Devon; not but what the Gannet can stand any amount of rough sea; and I have seen it dash after fish into seas that one would have thought must have rolled it over and drowned it, especially as it rose to the surface gulping down its prey.

It is included in Professor Ansted's list, but only marked as occurring in Guernsey. There are three specimens, an adult and two young, in the Museum.

163. COMMON TERN. *Sterna fluviatilis*, Naumann. French, "Hirondelle de mer," "Pierre garin." The Common Tern is a regular but not numerous spring and autumn visitant to the Islands, some remaining to breed. I do not know that it breeds anywhere in Guern-

sey itself, but it may do so, for in the Vale in the summer of 1878 I saw more than one pair about the two bays, Grand Havre and L'Ancresse, all through the summer; some of them certainly seemed paired, but I never could find where their nests were; some of the others apparently were non-breeding birds, as they did not appear to be paired. These bays and along the coast near St. Samson were the only places in Guernsey itself that I saw the Terns; there were some also about Herm, but we could not find any nests there; but Mr. Howard Saunders and myself found a few pairs breeding on one of the rocky islands to the north of Herm; when we visited them on the 27th June, 1878, we only found four nests, two with two eggs each and two with only one egg each. Probably these were a second laying, the nests having been robbed, as had everything else on these Islands; there must have been more than four nests there really, as there were several pairs of birds about, but we could not find any other nests; these four were on the hard rocks, with little or no attempt at a real nest. This was the only one of the small rocky islands on which we found Terns breeding, though we searched every one of them that had any land above water at high tide; the others, of course, were useless. I had expected for some time that Common Terns did breed on some of these rocks, as I have an adult female in full breeding-plumage, which had been shot on the 29th June, 1877, near St. Samson's, which is only about three miles from these Islands, and which certainly showed signs of having been sitting; and Mr. Jago, the bird-stuffer, had one in full breeding-plumage, killed at Herm early in June, 1878; but several of the sailors about, and some friends of mine who were in the habit of visiting these islands occasionally, seemed very sceptical on the subject; but Mr. Howard Saunders and I quite settled the question by finding the eggs, and we also thoroughly identified the birds. The Common Tern seemed to be the only species of Tern breeding on the rocks; we certainly saw nothing else, and no Common Terns even, except on the one island on which we found the eggs. The autumnal visitants are mostly young birds of the year, some of them, of course, having been bred on the Islands and others merely wanderers from more distant breeding-stations. No young Terns appeared to have flown when I left the Islands at the end of July; at least, I saw none about, though there were several adults about both Grand Havre and L'Ancresse Bay. The same remark applies to

Herm, where my last visit to the shell-beach was on the 22nd of July, when I saw several adult Common Terns about, but no young ones with them; all these were probably birds which had been robbed of one or more clutches of eggs.

Professor Ansted includes the Common Tern in his list, but only marks it as occurring in Guernsey. There is one specimen in the Museum, a young bird of the year.

164. ARCTIC TERN. *Sterna macrura*, Naumann. French, "Hirondelle de mer arctique." [31] —The Arctic Tern is by no means so common in the Islands as the Common Tern, and is, as far as I can make out, only an occasional autumnal visitant, and then young birds of the year most frequently occur, as I have never seen a Guernsey specimen of an adult bird. I do not think it ever visits the Islands during the spring migration; I did not see one about the Vale in the summer of 1878, nor did Mr. Howard Saunders and myself recognise one when we visited the rocks to the north of Herm. It may, however, have occurred more frequently than is supposed, and been mistaken for the Common Tern, so it may be as well to point out the chief distinctions: these are the short tarsus of the Arctic Tern, which only measures 0.55 of an inch, whilst that of the Common Tern measures 0.7 of an inch; and the dark grey next to the shaft on the inner web of the primary quills of the Arctic Tern, which is much narrower than in those of the Common Tern. These two distinctions hold good at all ages and in all states of plumage; as to fully adult birds in breeding plumage there are other distinctions, the tail of the Arctic Tern being much longer in proportion to the wing than in the Common Tern, and the bill being nearly all red instead of tipped with horn-colour.

The Arctic Tern is not included in Professor Ansted's list, and there is no specimen at present in the Museum.

165. BLACK TERN. *Hydrochelidon nigra*, Linnaeus. French, "Guifette noire," "Hirondelle de mer ipouvantail." [32] —The Black

Tern is by no means a common visitant to the Islands, and only makes its appearance in the autumn, and then the generality of those that occur are young birds of the year. I have one specimen of a young bird killed at the Vrangue on the 1st October, 1876. It does not seem to occur at all on the spring migration; at least I have never heard of or seen a Channel Island specimen killed at that time of year. As this is a marsh-breeding Tern, it is not at all to be wondered at that it does not, at all events at present, remain to breed in the Islands, there being so few places suited to it, though it is possible that before the Braye du Valle was drained, and large salt marshes were in existence in that part of the Island, the Black Tern may have bred there. I can, however, find no direct evidence of its having done so, and therefore can look upon it as nothing but an occasional autumnal straggler.

The Black Tern is not included in Professor Ansted's list, and there is no specimen in the Museum. These are all the Terns I have been able to prove as having occurred in the Channel Islands, though it seems to me highly probable that others occur—as the Sandwich Tern, the Lesser Tern, and the Roseate Tern (especially if, as I have heard stated, it breeds in small numbers off the coast of Brittany). Professor Ansted includes the Lesser Tern in his list, but that may have been a mistake, as my skin of a young Black Tern was sent to me for a Lesser Tern.

166. KITTIWAKE. *Rissa tridactyla*, Linnaeus. French, "Mouette tridactyle."—The Kittiwake is a regular and numerous autumn and winter visitant to all the Islands, sometimes remaining till late in the spring, which misled me when I made the statement in the 'Zoologist' for 1866 that it did breed in the Channel Islands; subsequent experience, however, has convinced me that the Kittiwake does not breed in any of the Islands. Captain Hubback, however, informed me that a few were breeding on the rocks to the south of Alderney in 1878, but when Mr. Howard Saunders and I went with him to the spot on the 25th June, we found no Kittiwakes there, all those Captain Hubback had previously seen having probably departed to their breeding-stations before our visit, and after they had been seen

by him some time in May. Professor Ansted includes the Kittiwake in his list, but only marks it as occurring in Guernsey and Sark. There are two specimens in the Museum, an adult bird and a young one in that state of plumage in which it is the Tarrock of Bewick and some of the older authors.

167. HERRING GULL. *Larus argentatus*, Gmelin. French, "Goeland argenti," "Goeland ` manteau bleu."—The Herring Gull is very common, indeed the commonest Gull, and is resident in all the Islands throughout the year, breeding in nearly all of them in such places as are suited to it. In Guernsey it breeds on the high cliffs, from the so-called Gull Cliff, near Pleinmont, to the Corbiere, the Gouffre, the Moye Point to Petit Bo in considerable numbers; from Petit Bo Bay to St. Martin's Point much more sparingly. In Sark it breeds in considerable numbers; on Little Sark on both sides of the Coupie, and on nearly all the west side; that towards Guernsey, especially about Harbour Goslin, a place called the Moye de Moutton near there, which is a most excellent place for watching the breeding operations of this Gull as well as of the Shags, as with a moderate climb on the rocks one can easily look into several nests and see what both old and young are about. On the island close to Sark, called Isle de Merchant, or Brechou, especially on the steep rocky side nearest to Sark; a great many also breed on and about the Autelets: in fact, almost all the grandest and wildest scenery in Sark has been appropriated by the Herring Gulls for their breeding-places, who, except for the Shags, hold almost undisputed possession of the grandest part of the Island. On the east side, or that towards France, few or no Herring Gulls breed; the cliffs being more sloping, and covered with grass and gorse, and heather, are not at all suited for breeding purposes for the Herring Gull. A few pairs have lately set up a small breeding-station on the rock before mentioned near Jethou, as La Fauconnihre; a very few also breed on Herm on the south part nearest to Jethou, but none that we could see on the rocks to the north of Herm. A great many breed also in Alderney on the south and east sides, but none on the little island of Burhou, which has been entirely appropriated by the Lesser Blackbacks; in all these places the Herring Gulls and Shags take almost

entire possession of the rocks, the Lesser Black-backs apparently never mixing with them; indeed, except a chance straggler or two passing by, a Lesser Black-back is scarcely to be seen at any of these stations. The Herring Gull and the Lesser Black-back, though very distinct in their adult plumage, and even before they fully arrive at maturity, as soon as they begin to show the different colour of the mantle, which they do in their second autumn, when a few of either the dark or the pale grey feathers appear amongst the brownish ones of the young bird, are before this change begins very much alike. In the down I think they are almost, if not quite, indistinguishable after that in their first feathers, and up to their first winter they appear to me distinguishable. As far as the primary quills go I do not see much difference; the shafts, perhaps, of the quills of the Lesser Black-back are darker than those of the Herring, but the difference if anything is very slight; but the head and neck and the centres of the feathers of the back of the Lesser Black-back are darker, — more of a dark smoky brown than those of the Herring Gull: this difference of colour is even more apparent on the under surface, including the breast, belly, and flanks. The shoulder of the wing and the under wing-coverts of the Lesser Black-back are much darker, nearly dull sooty black, and much less margined and marked with pale whitey brown than those of the Herring Gull. The dark bands on the end of the tail-feathers of the Lesser Black-back are broader and darker than in the Herring Gull: this seems especially apparent on the two outer tail-feathers on each side; besides this, there is a slight difference in the colour of the legs, those of the Lesser Black-back showing a slight indication of the yellow of maturity. I have noted these distinctions both from living specimens of both species which I have kept, and noted their various changes from time to time, and from skins of both: unfortunately the two skins of the youngest birds I have are not quite of the same age, one being that of a young Herring Gull, killed at the Needles in August, — the other a young Lesser Black-back, killed in Guernsey in December; but I do not think that this difference of time from August to December, the birds being of the same year, makes much difference in the colour of the feathers; at least this is my experience of live birds: it is not till the next moult that more material distinctions begin to appear; after that there can be no doubt as to the species. Two young Herring Gulls which I have, and which I saw in the flesh at Couch's shop just after they had been shot, seem to me worthy of some notice as showing the gradual change of plumage in the Herring Gull; they were shot on the same day, and appear to me to be one exactly a year older than the other; they were killed in November, when both had clean moulted, and show examples of the second and third moult. No. 1, the oldest, has the back nearly uniform grey, and the rump and upper tail-coverts white, as in the adult. In No. 2, the younger one, the grey feathers on the back were much mixed with the brownish feathers of the young bird, and there are no absolutely white feathers on the rump and tail-coverts, all of them being more or less marked with brown. The tail in No. 2 has the brown on it collected in large and nearly confluent blotches, whilst that of No. 1 is merely freckled with brown. But perhaps the greatest difference is in the primary quills; the first four primaries, however, are much alike, those of No. 1, being a little darker and more distinctly coloured; in both they are nearly of a uniform colour, only being slightly mottled on the inner web towards the base; there is no white tip to either. In No. 1 the fifth primary has a distinct white tip; the sixth also has a decided white tip, and is much whiter towards the base, the difference being quite as perceptible on the outer as on the inner web. The seventh has a small spot of brown towards the tip on the outer web, the rest of the feather being almost uniform pale grey, with a slightly darker shade on the outer web, and white at the tip; the eighth grey, with a broad white tip. In No. 2 the fifth primary has no white tip; the sixth also has no white tip, and not so much white towards the base; the seventh is all brown, slightly mottled towards the base, and only a very slight indication of a white tip; and the eighth is mottled throughout. I think it worth while to mention these two birds, as I have their exact dates, and the difference of a year between them agrees exactly with young birds which I have taken in their first feathers and brought up tame. I may also add, with regard to change of plumage owing to age, that very old birds do not appear to get their heads so much streaked with brown in the winter as younger though still adult birds, as a pair which I caught in Sark when only flappers, and brought home in July, 1866, had few or no brown streaks about their heads in the winter of 1877-8, and in the winter of 1878-9 their heads are almost as white as in the breeding-season. These birds had their first brood in 1873, and have bred regularly every year since that time, and certainly have considerably more white on their primary quills than when they first assumed adult plumage and began to breed. Probably this increase of white on the primaries as age increases, even after the full-breeding-plumage is assumed, is always the case in the Herring Gull, and also in both the Lesser and Greater Black-backs, thus distinguishing very old birds from those which, though adult, have only recently assumed the breeding-plumage. I know Mr. Howard Saunders is of this opinion, certainly as far as Herring Gulls are concerned. Besides the live ones, two skins I have, both of adult birds, as far as breeding-plumage only is concerned, are evidently considerably older than the other. No. 1, the youngest of these, — shot in Guernsey in August, when just assuming winter plumage, the head being much streaked, even then, with brown, showing that though adult it was not a very old bird, — has the usual white tip on the first primary, below which the whole feather is black on both webs, and below that a white spot on both webs, for an inch; the white, however, much encroached upon on the outer part of the outer web by a margin of black. In No. 2, probably the older bird, the first primary has the white tip and the white spot running into each other, thus making the tip of the feather for nearly two inches white, with only a slight patch of black on the outer web. On the second primary of No. 1 the white tip is present, but no white spot; but on the same feather of No. 2 there is a white spot on the inner web, about an inch from the white tip; this would, probably, in a still older bird, become confluent with the white tip, as in the first primary. I have not, however, a sufficiently old bird to follow out this for certain. In No. 1, the older bird, the pale grey on the lower part of the feathers also extends farther towards the tip, thus encroaching on the black of the primaries from below as well as from above. I think these examples are sufficient to show that the white does encroach on the black of the primaries as the bird grows older, till at last, in very old birds, there would not be much more than a bar of black between the white tip and the rest of the feather; and this is very much the case with the tame ones I caught in Sark in 1866, and which are therefore, now in the winter of 1879, twelve and a half years old; but I do not believe that at any age the black wholly disappears from the primaries, leaving them white as in the Iceland and Glaucous Gulls. The Herring Gull is an extremely voracious bird, eating nearly everything that comes in its way, and rejecting the indigestible parts as Hawks do. Mr. Couch, in the 'Zoologist' for 1874, mentions having taken a Misseltoe Thrush from the throat of one; and I can quite believe it, supposing it found the Thrush dead or floating half drowned on the water. I have seen my tame ones catch and kill a nearly full-grown rat, and bolt it whole; and young ducks, I am sorry to say, disappear down their throats in no time, down and all. They are also great robbers of eggs, no sort of egg coming amiss to them; Guillemots' eggs, especially, they are very fond of; this may probably account for there being no Guillemots breeding in Guernsey or Sark, and only a very few at Alderney; in fact, Ortack being the only place in the Channel Islands in which they do breed in anything like numbers.

Professor Ansted includes the Herring Gull in his list, but only marks it as occurring in Guernsey and Sark. There are two, an old and a young bird, in the Museum.

168. LESSER BLACK-BACKED GULL. *Larus fuscus*, Linnaeus. French, "Goeland ` pieds jaunes."—The Lesser Black-backed Gull is common in the Islands, remaining throughout the year and breeding in certain places. None of these birds breed in Guernsey itself, or on the mainland of Sark, and very few, if any, on Alderney. A few may be seen, from time to time, wandering about all the Islands during the breeding-season; but these are either immature birds or wanderers from their own breeding-stations. About Sark a few pairs breed on Le Tas [33] and one or two other outlying islets; their principal breeding-stations, however, appear to be on the small rocky islands to the north of Herm, on all of which, as far out as the Amfrocques, we found considerable numbers breeding, or rather attempting to do so; for this summer, 1878, having been generally fine, all these rocks were tolerably easily landed on, and the fishermen had robbed the Lesser Black-backs to an extent which threatens some day to exterminate them, in spite of the Guernsey Bird Act, which professes to protect the eggs as well as the birds; but a far better protection for these poor Black-backs is a roughish summer, when landing on these islands is by no means safe or pleasant, and frequently impossible. On Burhou, near Alderney, there are also a

considerable number of Lesser Black-backs breeding, though they fare quite as badly from the Alderney and French fishermen as those on the Amfrocques and other islands north of them do from the Guernsey fishermen. On all these islands the nests of the Lesser Black-backs were placed amongst the bracken, sea stock, thrift, &c, which grew amongst the rocks, and on the shallow soil which had collected in places. When I was at Burhou in 1876 I found Lesser Black-backs breeding all over the Island, some of the nests being placed on the low rocks, some amongst the bracken and thrift; so thickly scattered amongst the bracken were the nests, that one had to be very careful in walking for fear of treading on the nests and breaking the eggs. On this Island there is an old deserted cottage, sometimes used as a shelter by the lessees of the Island, who go over there to shoot a few wretched rabbits which pick up a precarious subsistence by feeding on the scanty herbage; on the roof of this cottage several of the Lesser Black-backs perched themselves in a row whilst I was looking about at the eggs, and kept up a most dismal screaming at the top of their voices. The eggs, as is generally the case with gulls, varied considerably both in ground colour and marking; some were freckled all over with small spots—dark brown, purple, or black; others had larger markings, principally collected at the larger end; the ground colour was generally blue, green, or dull olive-green. None of the Gulls had hatched when I was there on the 14th of June, though some of the eggs were very hard set; and on the 29th of July I received two young birds which had been taken on Burhou; these still had down on them when I got them, and were then difficult to tell from young Herring Gulls. The distinctions I have mentioned in my note of that bird were, however, apparent, and the slight difference in the colour of the legs is perhaps more easily seen in the live birds than in skins which have been kept and faded into "Museum colour." It is some time, however, before either bird assumes the proper colour, either of the legs or bill, the change being very gradual. After the autumnal moult of 1878, however, the dark feathers of the mantle almost entirely took the place of the brownish feathers of the young birds; the quills, however, have still (February, 1879) no white tips, and the tail-feathers are still much mottled with brown. One Lesser Black-back, which I shot near the Vale Church on the 17th of July, 1866, is perhaps worthy of note as being in transition, and perhaps a rather abnormal state of change considering the time of year at which it was shot; it was in a full state of moult; the new feathers on the head, neck, tail-coverts, and under parts are white; the tail also is white, except four old feathers, two on each side not yet moulted, which are much mottled with brown. The primary quills had not been moulted, and are quite those of the immature bird, with no white tip whatever. All the new feathers of the back and wing-coverts are the dark slate-grey of the adult, but the old worn feathers are the brownish feathers of the young bird; these feathers are much worn and faded, being a paler brown than is usual in young birds. The legs and bill are also quite as much in a state of change as the rest of the bird. Before finishing this notice of the Lesser Blackback I think it is worth while to notice that it selects quite a different sort of breeding-place to the Herring Gull; the nests are never placed on ledges on the steep precipitous face of the cliffs, but amongst the bracken and the flat rocks, as at Burhou, the only rather steep rock I have seen any nests on was at the Amfrocques, but there they were on the flattish top of the rock, and not on ledges on the side.

Professor Ansted includes the Lesser Black-backed Gull in his list, but only marks it as occurring in Guernsey. There is one specimen in the Museum.

169. COMMON GULL. *Larus canus*, Linnaeus. French, "Goeland cendri," "Mouette a pieds bleus," [34] "La Mouette d'Hiver". [35] — The Common Gull, though by no means uncommon in the Channel Islands during the winter, never remains to breed there, nor does it do so, I believe, any where in the West of England, certainly not in Somerset or Devon, as stated by Mr. Dresser in the 'Birds of Europe,' *fide* the Rev. M.A. Mathew and Mr. W.D. Crotch, who must have made some mistake as to its breeding in those two counties; in Cornwall it is said to breed, by Mr. Dresser, on the authority of Mr. Rodd. Mr. Dresser, however, does not seem to have had his authority direct from either of these gentlemen, and only quotes it from Mr. A.G. More. Mr. Rodd, however, in his 'Notes on the Birds of Cornwall,' published in the 'Zoologist' for 1870, only says, "Generally

distributed in larger or smaller numbers along or near our coasts," which would be equally true of the Channel Islands, although it does not breed there; however, as Mr. Rodd is going to publish his interesting notes on the Birds of Cornwall in a separate form, it is much to be hoped that he will clear that matter up as far as regards that county and the Scilly Islands. Like the Herring and Lesser Black-backed Gull, the Common Gull goes through several changes of plumage before it arrives at maturity; like them it begins with the mottled brownish stage, and gradually assumes the blue-grey mantle of maturity; in the earlier stages the primaries have no white spots at the tips. The legs and bill, which appear to go through more changes than in other Gulls, are in an intermediate state bluish grey (which accounts for Temminck's name mentioned above) before they assume the pale yellow of maturity: although at this time they have the mantle quite as in the adult, there is a material difference in the pattern of the primary quills, and they do not appear to breed till their bills have become quite yellow and their legs a pale greenish yellow. I cannot quite tell at what age the Common Gull begins to breed, for, although I have a pair which have laid regularly for the last two years (they have not, however, hatched any young, which perhaps is the fault of the Herring Gulls, whom I have several times caught sucking their eggs), I do not know what their age was when I first had them as I did the Herring Gulls from Sark and the Lesser Black-backs from Burhou; I can only say when I first had them they had the bills and legs blue; in fact they were in the state in which they are the "Mouette ` pieds bleus" of Temminck.

Professor Ansted includes the Common Gull in his list, and marks it as occurring in Guernsey and Sark. There is no specimen in the Museum.

170. GREAT BLACK-BACKED GULL. *Larus marinus*, Linnaeus. French, "Goeland ` manteau noir." — The Great Black-backed Gull is by no means so numerous in the Channel Islands as the Herring Gull and the Lesser Black-back, and is here as elsewhere a rather solitary and roaming bird. A few, however, remain about the Channel Islands, and breed in places which suit them, such as Ortack,

which I have before mentioned, as the breeding-place of the Razorbill and Guillemot; and we found one nest on one of the rocks to the north of Herm, but it had been robbed, as had all the other Gulls' nests about there; we saw, however, the old birds about, and Mr. Howard Saunders found one nest on the little Island of Le Tas, close to Sark; it was quite on the top of the Island, and there were young in it. I have one splendid adult bird, shot near the harbour in Guernsey, in March: I should think this is rather an old bird, as, although there are slight indications of winter plumage on the head, the white tips of the primaries are very large, that of the first extending fully two inches and a half, which is considerably more than that of a fully adult bird I have from Lundy Island. The Great Black-backed Gull is sufficiently common and well known to have a local name in Guernsey-French (Hublot or Ublat), for which see 'Mitivier's Dictionary.'

Professor Ansted includes the Great Black-backed Gull in his list, and marks it as only occurring in Guernsey and Sark. There are three specimens in the Museum—an adult bird, a young one, and a young one in down, with the feathers just beginning to show. In the young bird the head and neck were mottled and much like those of a young Herring Gull in the same state; the back, thighs, and under parts do not appear so much spotted as in the young Herring Gull; the feathers on the scapulars and wing-coverts were just beginning to show two shades of brown, as in the more mature state; the same may be said of the primary quills, which were also just beginning to make their appearance; the tail, which was only just beginning to show, was nearly black, margined with white.

171. BROWN-HEADED GULL. *Larus ridibundus*, Linnaeus. French, "Mouette rieuse." [36] This pretty little Gull is a common autumn and winter visitant to all the Islands, remaining on to the spring, but never breeding in any of them, though a few young and non-breeding birds may be seen about at all times of the summer, especially about the harbour. Being a marsh-breeding Gull, and selecting low marshy islands situated for the most part in inland fresh-water lakes and large pieces of water, it is not to be wondered

at that it does not breed in the Channel Islands, where there are no places either suited to its requirements or where it could find a sufficient supply of its customary food during the breeding-season. Very soon after they have left their breeding-stations, however, both old and young birds may be seen about the harbours and bays of Guernsey and the other islands seeking for food, in which matter they are not very particular, picking up any floating rubbish or nastiness they may find in the harbour. The generality of specimens occurring in the Channel Islands are in either winter or immature plumage, very few having assumed the dark-coloured head which marks the breeding plumage. This dark colour of the head, which is sometimes assumed as early as the end of February, comes on very rapidly, not being the effect of moult, but of a change of colour in the feathers themselves, the dark colouring-matter gradually spreading over each feather and supplanting the white of the winter plumage; a few new feathers are also grown at this time to replace any that have been accidentally shed — these come in the dark colour. The young birds in their first feathers are nearly brown, but the grey feathers make their appearance amongst the brown ones at an earlier stage than in most other gulls. The primary quills, which are white in the centre with a margin of black, vary also a good deal with age, the black margins growing narrower and the white in places extending through the black margin to the edge, so that in adult birds the black margins are not so complete as in younger examples.

Professor Ansted mentions the Laughing Gull in his list, by which I presume he means the present species, and marks it as only occurring in Guernsey. There is no specimen in the Museum. As it is just possible that the Mediterranean Black-headed Gull, *Larus melanocephalus,* may occur in the Islands, — as it does so in France as far as Bordeaux, and has once certainly extended its wanderings as far as the British Islands, — it may be worth while to point out the principal distinctions. In the adult bird the head of *L. melanocephalus* in the breeding-season is black, not brown as in *L. ridibundus*, and the first three primaries are white with the exception of a narrow streak of black on the outer web of the first, and not white with a black margin as in *L. ridibundus*. In younger birds, however, the primaries are a little more alike, but the first primary of *L. melanocephalus* is black or nearly so; in this state Mr. Howard Saunders has given plates of the first three primaries of *L. melanocephalus* and *L. ridibundus*, both being from birds of the year shot about March, in his paper on the *Larinae*, published in the 'Proceedings of the Zoological Society' for the year 1878.

172. LITTLE GULL. *Larus minutus*, Pallas. French, "Mouette pygmie."—I have never met with this bird myself in the Channel Islands, nor have I seen a Channel Island specimen, but Mr. Harvie Brown, writing to the 'Zoologist' from St. Peter's Port, Guernsey, under date January 25th, says, "In the bird-stuffer's shop here I saw a Little Gull in the flesh, which had been shot a few days ago." [37] Mr. Harvie Brown does not give us any more information on the subject, and does not even say whether the bird was a young bird or an adult in winter plumage; but probably it was a young bird of the year in that sort of young Kittiwake or Tarrock plumage in which it occasionally occurs on the south coast of Devon.

Professor Ansted does not include the Little Gull in his list, and there is no specimen in the Museum.

173. GREAT SHEARWATER. *Puffinus major*, Faber. French, "Puffin majeur." [38] —I think I may fairly include the Great Shearwater in my list as an occasional wanderer to the Islands, as, although I have not a Channel Island specimen, nor have I seen it near the shore or in any of the bays, I did see a small flock of four or five of these birds in July, 1866, when crossing from Guernsey to Torquay. We were certainly more than the Admiralty three miles from the land; but had scarcely lost sight of Guernsey, and were well within sight of the Caskets, when we fell in with the Shearwaters. They accompanied the steamer for some little way, at times flying close up, and I had an excellent opportunity of watching them both with and without my glass, and have therefore no doubt of the species. There was a heavyish sea at the time, and the Shearwaters were generally flying under the lee of the waves, just rising sufficiently to avoid the

crest of the wave when it broke. They flew with the greatest possible ease, and seemed as if no sea or gale of wind would hurt them; they never got touched by the breaking sea, but just as it appeared curling over them they rose out of danger and skimmed over the crest; they never whilst I was watching them actually settled on the water, though now and then they dropped their legs just touching the water with their feet.

The Great Shearwater is not mentioned in Professor Ansted's list, and there is no specimen in the Museum.

174. MANX SHEARWATER. *Puffinus anglorum*, Temminck. French, "Petrel Manks."—The Manx Shearwater can only be considered as an occasional wanderer to the Channel Islands, and never by any means so common as it is sometimes on the opposite side of the Channel about Torbay, especially in the early autumn. I have one Guernsey specimen, however, killed near St. Samson's on the 28th September, 1876. [39] As far as I can make out the Manx Shearwater does not breed in any part of the Channel Islands, but being rather of nocturnal habits at its breeding-stations, and remaining in the holes and under the rocks where its eggs are during the day, it may not have been seen during the breeding-season; but did it breed anywhere in the Islands more birds, both old and young, would be seen about in the early autumn when the young first begin to leave their nests; and the Barbelotters would occasionally come across eggs and young birds when digging for Puffins' eggs.

The Manx Shearwater is not included in Professor Ansted's list, and there is no specimen in the Museum.

175. FULMAR PETREL. *Fulmarus glacialis*, Linnaeus. French, "Petrel fulmar."—The Fulmar Petrel, wandering bird as it is, especially during the autumn, at which time of year it has occurred in all the western counties of England, very seldom finds its way to the Channel Islands, as the only occurrence of which I am aware is one which I picked up dead on the shore in Cobo Bay on the 14th of

November, 1875, after a very heavy gale. In very bad weather, and after long-continued gales, this bird seems to be occasionally driven ashore, either owing to starvation or from getting caught in the crest of a wave when trying to hover close over it, after the manner of a Shearwater, as this is the second I have picked up under nearly the same circumstances, the first being in November, 1866, when I found one not quite dead on the shore near Dawlish, in South Devon. It must be very seldom, however, that the Fulmar visits the Channel Islands, as neither Mr. Couch nor Mrs. Jago had ever had one through their hands, and Mr. MacCulloch has never heard of a Channel Island specimen occurring.

It is not included in Professor Ansted's list, and there is no specimen in the Museum.

176. STORM PETREL. *Thalassidroma pelagica* Linnaeus. French, "Thalassidrome tempjte."—Mr. Gallienne, in his remarks published with Professor Ansted's list, says, "The Storm Petrel breeds in large numbers in Burhou, a few on the other rocks near Alderney, and occasionally on the rocks near Herm; these are the only places where they breed, although seen and occasionally killed in all the Islands." I can add to these places mentioned by Mr. Gallienne the little island, frequently mentioned before, near Sark, Le Tas, where Mr. Howard Saunders found several breeding on the 24th June, 1878. I could not accompany him on this expedition, so he alone has the honour of adding Le Tas to the breeding-places of the Storm Petrel in the Channel Islands, and he very kindly gave me the two eggs which he took on that occasion. When I visited Burhou in June, 1876, I was unsuccessful in finding more than part of a broken egg and a wing of a dead bird. But Colonel L'Estrange, who had been there about a fortnight before, found two addled eggs, but saw no birds. I thought at the time that I had been too late and the birds had departed, but this does not seem to have been the case, as Captain Hubback wrote to me in July of this year (1878), and said, "Do you not think that perhaps you were early on the 14th of June? Of the six eggs I took on the 2nd of July this year, two were quite fresh, three hard-sat, and one deserted." I have no doubt he was right, as

the wing of the dead bird I found was, no doubt, that of one that had come to grief the year before, and the egg was one which had been sat on and hatched, and might therefore have been one of the previous year; and the same, possibly, might have been the case with Col. L'Estrange's two addled eggs. It appears, however, to be rather irregular in its breeding habits, nesting from the end of May to July or August. In Burhou the Storm Petrel bred mostly in holes in the soft black mould, which was also partly occupied by Puffins and Babbits, but occasionally under large stones and rocks. We did not find any breeding on the islands to the north of Herm, but they may do so occasionally, in which case their eggs would probably be mostly placed under large rocks and stones, where the Puffins find safety from the attacks of the various egg-stealers. At other times of year than the breeding-season, the Storm Petrel can only be considered an occasional storm-driven visitant to the Islands.

It is included in Professor Ansted's list, and marked as occurring in Alderney, Sark, Jethou, and Herm.

With this bird ends my list of the Birds of Guernsey and the neighbouring Islands. It contains notices of only 176 birds, 21 less than Professor Ansted's list, which contains 197; but it seems to me very doubtful whether many of these 21 species have occurred in the Islands. I can find no other evidence of their having done so than the mere mention of the names in that list, as, except the few mentioned in Mr. Gallienne's notes, no evidence whatever is given of the when and where of their occurrence; and we are not even told who was responsible for the identification of any of the birds mentioned. I have no doubt, however, that any one resident in the Islands for some years, and taking an interest in the ornithology of the district, would be able to add considerably to my list, as Miss C.B. Carey, had she lived, would no doubt have enabled me to do. I think it very probable, mine having been only flying visits, though extending over several years and at various times of year, I may have omitted some birds, especially amongst the smaller Warblers and the Pipits, and perhaps amongst the occasional Waders. There is one small family—the Skuas—entirely unrepresented in my list; I am rather surprised at this as some of them, especially the Pomatorhine—or, as it is perhaps better known, the Pomerine—Skua, *Stercorarius pomatorhinus*, and Richardson's Skua, *Stercorarius crepi-*

datus, are by no means uncommon on the other side of the Channel, about Torbay, during the autumnal migration; but I have never seen either species in the Island, nor have I seen a Channel Island skin, nor can I find that either the bird-stuffers or the fishermen and the various shooters know anything about them. I have therefore, though I think it by no means; unlikely that both birds occasionally occur, thought it better to omit their names from my list.

Professor Ansted has only mentioned one of the family—the Great Skua, *Stercorarius catarrhactes*,—in his list, which also may occasionally occur, as may Buffon's Skua, *Stercorarius parasiticus*; but neither of these seem to me so likely to occur as the two first-mentioned, not being by any means so common on the English side of the Channel.

In bringing my labours to a conclusion I must again thank Mr. MacCulloch and others, who have assisted me in my work either by notes or by helping in out-door work.

FINIS.

ENDNOTES

[1]

a Alderney.
e Guernsey.
i Jersey.
o Sark.
u Jethou and Herm.

[2]

This was nearly the whole of the Vale, including L'Ancresse Common.

[3]

Fourteen "livres tournois" are about equal to #1.

[4]

This Act is passed annually at the Chief Pleas after Easter.

[5]

Falco aesalon, Tunstall, H.S. 1771. *Falco aesalon*, Gmelin, Y., 1788.

[6]

See Temminok.

[7]

See 'Birds of Spain,' by Howard Saunders, Esq., published in the works of the Sociiti Zoologique de France, where he says: — "C. *ceruginosus* et C. *cyaneus* ont les lisihres extirieures des remiges imarginies, jusqu'` et y comprise la cinquihme, et cette forme se trouve en presque toutes les *Circus* exotiques. En C. *swainsonii* (the Pallid Harrier) et C. *cineraceus* cette imargination successive se borne a la quatrieme." We have little to do with this distinction, except as between C. *cyaneus* and C. *cineraceus*, C. *aeruginosus* being otherwise sufficiently distinct, and C. *swainsonii* not coming within our limits.

[8]

"Tereus," I soon found, as I expected, was Mr. MacCulloch.

[9]

These reeds are the common reed Spires, Spire-reed, or Pool-reed. *Arundo phragmites*. See 'Popular Names of British Plants,' by Dr. Prior, p. 219.

[10]

This name of Temminck is no doubt applied to the Continental form, *Acredula caudata*, of Linnaeus, not to the British form now elevated into a species under the name *Acredula rosea*, of Blyth. Owing to want of specimens I have not been able to say to which form the Channel Island Long-tailed Tit belongs, probably supposing them to be really distinct from *A. rosea*. *A. caudata* may, however, also occur, as both forms do occasionally, in the British Islands.

[11]

See Temminck's 'Man. d'Ornith.'

[12]

Dresser's 'Birds of Europe,' *fide* Degland's Grebe.

[13]

Where both forms are common this constantly happens — indeed, so constantly that Professor Newton, in his new edition of 'Yarrell,' has made but one species of the Black Crow and the Grey or Hooded Crow, *Corvus corone* and *Corvus cornix*, on the several grounds that there is no structural difference between the two; that their habits, food, cries, and mode of nidification are the same (in considering this, of course both forms must be traced throughout the whole of their geographical range, and not merely through the British Islands); that their geographical distribution is sufficiently similar not to present any difficulty; that they breed freely together; and that their offsprings are fertile, a very important consideration in judging whether two forms should be separated or joined as one species. This last seems to me to present the greatest difficulty, and the evidence at present appears scarcely conclusive. Of course in the limits of a note to a work like the present it is impossible to discuss so large a question. I can only refer my readers to Professor Newton's work, where they will find nearly all that can be said on the subject, and the reasons which have induced him to come to the conclusion he has.

[14]

Rim. Gu., p. 35.

[15]

Query, was this done by a migratory flock, as peas would be ripe about June or July, when migratory flocks of Wood Pigeons would not be likely to occur; or was the damage to newly sown peas in the spring?

[16]

For one instance see notice of the Quail; and the bird-stuffer had several other eggs besides those in the same nest as the Quails.

[17]

Fide Mr. MacCulloch.

[18]

See 'Dresser's Birds of Europe.'

[19]

For the last, see Temminck's 'Man, d'Ornithologie.'

[20]

See 'Zoologist' for 1867, p. 829.

[21]

Temminck, 'Man. d'Ornithologie.'

[22]

See Temminck, 'Man. d'Ornithologie.'

[23]

The one above mentioned.

[24]

See 'Zoologist' for 1870, p. 2244.

[25]

"Hucard" in Guernsey French (see 'Metevier's Dictionary,') who also says "Notre Hucard est le Whistling Swan ou Hooper des Anglais."

[26]

See Temminck's 'Man. d'Ornithologie.'

[27]

See also Mitivier's Dictionary.

[28]

See note in 'Zoologist' for 1866.

[29]

'De la Mue du Bec et des Ornements Palpibraux du Macareux Arctique aprhs la Saison des Amours.' Par le Docteur Louis Bureau; 'Bulletin de la Sociiti Zoologique de France.'

[30]

'Zoologist' for 1869.

[31]

See Temininck, 'Man. d'Ornithologie.'

[32]

Temminck, 'Man. d'Ornithologie.'

[33]

Le Tas is often written L'Etat, but, as Professor Ansted says, "There can be no doubt it alludes to the form of the rock, viz., 'Tas,' a heap such as is made with hay or corn."

[34]

See Temminck's 'Man. d'Ornithologie.'

[35]

Buffon.

[36]

See Temminck's 'Man. d'Ornithologie.'

[37]

See 'Zoologist' for 1869, p. 1560.

[38]

See Temminck, 'Man. d'Ornithologie.'

[39]

This is since my note to Mr. Dresser, published in his 'Birds of Europe,' when I said I had never seen it in the Channel Islands, although it probably occasionally occurred there.